A. W. (Arthur William) Poyser

Magnetism and Electricity

A. W. (Arthur William) Poyser

Magnetism and Electricity

ISBN/EAN: 9783744791076

Printed in Europe, USA, Canada, Australia, Japan

Cover: Foto ©berggeist007 / pixelio.de

More available books at **www.hansebooks.com**

MAGNETISM AND ELECTRICITY

PRINTED BY
SPOTTISWOODE AND CO., NEW-STREET SQUARE
LONDON

BY

ARTHUR WM. POYSER, M.A.

TRINITY COLLEGE, DUBLIN

ASSISTANT MASTER IN THE WYGGESTON AND QUEEN ELIZABETH'S
GRAMMAR SCHOOL, LEICESTER

LONDON

LONGMANS, GREEN, AND CO.

AND NEW YORK : 15 EAST 16th STREET

1889

PREFACE

THIS SMALL VOLUME is intended as an introduction for beginners, and primarily for those who are reading for the South Kensington Elementary Examination in Magnetism and Electricity. It will be seen, however, that several chapters which are not necessary for that examination have been inserted in order to cover the course usually taken in a year's school work.

The book, which is by no means an examination manual only, is the result of practical experience in teaching, and it has been thrown into experimental form from a conviction that, if the student is to gain an adequate knowledge of the subject, it is absolutely necessary for him to acquire it by experiment. Scientific knowledge derived from mere book-work, with a view to pass some particular examination, is almost useless, and indeed is not unlikely to produce a result the opposite of that intended by the student.

Of the two hundred and thirty-five illustrations, many have been drawn specially for this work from apparatus in common use in the laboratory of this school; others have been taken from Ganot's 'Physics' and Ganot's 'Popular Natural Philosophy;' and a few have been adapted from Professor Tyndall's 'Lessons in Electricity.' It is hoped that they may be used solely as aids in understanding the text, and not as substitutes for experimental work.

The exercises which are interspersed throughout the book

include questions from the South Kensington Examination, from the London University Matriculation Examination, and from those used in my own school lessons. Numerical examples are frequently given, as, even in the most elementary work, the student should learn that some knowledge of mathematics is not only useful but essential.

I must acknowledge my obligation to several friends and colleagues for their help and suggestions, particularly to the Rev. EDWARD ATKINS, B.Sc., who has given me the benefit of his long experience as a teacher, and to Mr. ALBAN JAMES, M.A., to whom I am indebted for valuable assistance in passing the work through the press.

A. W. P.

THE WYGGESTON SCHOOL, LEICESTER.
    August 1889.

CONTENTS

————

MAGNETISM.

FRICTIONAL ELECTRICITY.

MAGNETISM.

—◦◆◦—

CHAPTER I.

MAGNETIC ATTRACTION AND REPULSION.

Natural Magnets.—Exp. 1. Examine a piece of lodestone. Notice that it is a hard, dark-coloured, stone-like body. Dip it into iron filings; on withdrawal observe that the filings cling in tufts to certain parts (fig. 1).

FIG. 1.

Lodestone (although not always magnetic) is widely distributed in nature, being particularly abundant in Norway and Sweden, and in some parts of America. It was, however, originally found in *Magnesia*, in Asia Minor; from this circumstance probably it was called *magnēs* by the Greeks, whence we have derived our words *magnet, magnetism,* &c. Since this substance is found in a natural state it is called a **natural magnet.** It is an iron ore, consisting of iron and oxygen, having the chemical formula Fe_3O_4, and is commonly called *magnetic oxide of iron.*

Exp. 2. Suspend a piece of lodestone (which has been shaped with a hammer in such a way that its attractive power is most apparent near the two ends) so that it can turn freely. This may be done in several ways; one of them is to

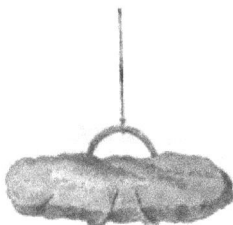

FIG. 2.

bend a piece of wire into the shape shown in fig. 2. This is then fastened with a piece of raw silk to a support (a horizontal gas-

B

bracket answers well for this purpose). Allow the lodestone to become steady, and notice that it points nearly north and south. If disturbed from this position it oscillates for a time, but then comes to rest in exactly the same position as at first, and, moreover, the same end always points in the same direction.

On account of this remarkable property, and its subsequent use in navigation, this substance is called **Lodestone**, from the Anglo-Saxon word *loedan*, to lead.

Exp. 3. Take an ordinary steel knitting-needle and dip it into iron filings. The filings do not cling to the needle.

Exp. 4. Draw the needle several times over a lodestone, taking care to move it always in the same direction, not to and fro. Again plunge it into the iron filings, and on withdrawal notice the tufts, similar to those in Exp. 1 clinging near the ends.

Exp. 5. Suspend a rod of iron, a rod of wood, and the needle used in the last experiment in three paper stirrups. (A paper

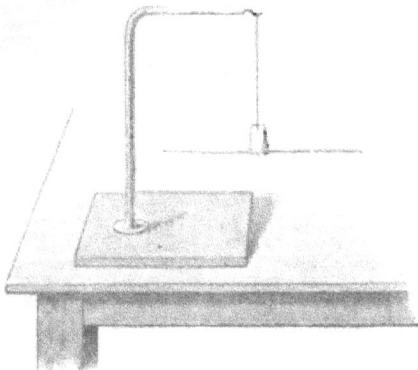

FIG. 3.

stirrup is made by doubling a strip of paper about two inches long, and fastening the free ends by a thread of untwisted silk; the other end of the silk is then tied to a suitable support, fig. 3.) Observe that the needle alone sets itself in a north and south direction. Thus the results obtained with the needle in Exps. 4 and 5 are exactly similar to those obtained with the natural magnet in Exps. 1 and 2.

We therefore learn that by passing the needle over a piece of lodestone we have imparted a new property to it, which manifests itself in several ways; in fact, we have made an **artificial magnet.**[1]

The process by means of which a steel needle is rendered capable of attracting small pieces of iron, and of setting itself,

[1] Artificial magnets are, however, usually made by methods which will be described later.

when freely suspended, in a north and south direction, is called *magnetisation*. The body is said to be *magnetised*, or to possess magnetism.

For experimental purposes we find that artificial magnets can not only be made more powerful, but are more convenient than natural ones. The most common forms are : (1) the *bar magnet*, sometimes cylindrical, generally rectangular (fig. 4) ; and (2) the *horseshoe magnet*, of the shape shown in fig. 5.

Exp. 6. Hang a horseshoe magnet vertically by a thread fastened at the bend. Notice that it comes to rest in a north and south direction.

Fig. 4.

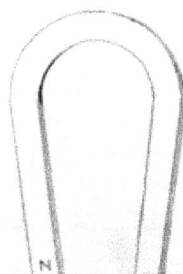

Fig. 5.

Attraction of Iron by Magnets.—Exp. 7. Take an ordinary bar magnet, and dip it into iron filings. Observe that the filings do

Fig. 6.

not adhere to all parts of the magnet, but that they accumulate in tufts near the ends (fig. 6).

Exp. 8. Take a strong bar magnet, and a number of small soft iron bars of equal size and weight.

(1) Near the end *a* of the magnet (fig. 7.) attach the greatest number of these bars that can be sustained. (2) Test a point *b* on

Fig. 7.

the magnet nearer the centre. It will be found that the same number of bars will no longer be supported. (3) Hang as many bars as possible at the point *b*. (4) At a point *c*, still nearer the centre, place an equal number to those sustained at the point *b*, and notice that some will fall. (5) Place a very small particle of iron—e.g., an iron filing—at the middle of the magnet; there is no attractive force whatever. (6) Test the other half of the magnet similarly; it will be found that equal weights are supported at equal distances from each end.

We therefore learn that—

(*a*) The attractive power of the magnet is greatest near the ends. Strictly speaking, there are two points, one near each end of the magnet, where the attractive power is greatest. These points are called the *poles* of the magnet.

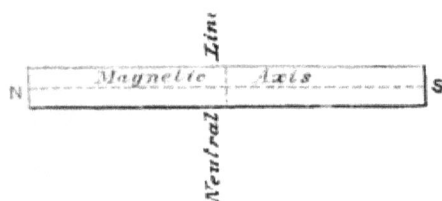

FIG. 8.

(*b*) The portion between the poles is apparently less magnetic.

(*c*) All round the magnet, midway between the poles, there is no attraction. This is called the *neutral line.*

The line joining the poles is called the *axis* (fig. 8).

Mutual Attraction between Magnets and Iron.—Exp. 9.

(1) Place a cork on the surface of water, and balance an iron rod on it (fig. 9). Bring a magnet near, and notice attraction.

FIG. 9.

(2) Replace the iron rod by a magnet, and present to it a piece of iron. Attraction ensues.

Exp. 10. The above experiment may be varied by placing a poker near the ends of a small horseshoe magnet (fig. 10). The magnet will be attracted by the poker.

It is well to mention that the mutual attraction between a magnet and a piece of iron, exhibited in these experiments, is altogether independent of the mutual attraction exerted by bodies in obedience to the law of gravitation. The latter law teaches us that all bodies mutually attract each other—e.g., the earth attracts the sun, the sun attracts the earth—while magnetic attraction can only be observed between certain substances

FIG. 10.

(to be afterwards mentioned), one or both of which must be magnetised.

EXERCISE I.

1. State what you know about the lodestone or natural magnet.
2. Give the derivation of the words *magnet* and *lodestone*.
3. Give experiments to show that the attraction between a magnet and iron is mutual.
4. What is meant by the terms *pole, neutral line, axis* of a magnet?
5. How would you roughly ascertain, by means of a magnetised needle, the position of the four cardinal points?

Action of Magnets on Magnets.—Let us now ascertain if there is any difference between the action of a magnet on (1) soft iron ; and (2) another magnet.

Exp. 11. Obtain or make a magnetic needle of the shape shown in fig. 11. It consists of a light strip of magnetised steel, fitted

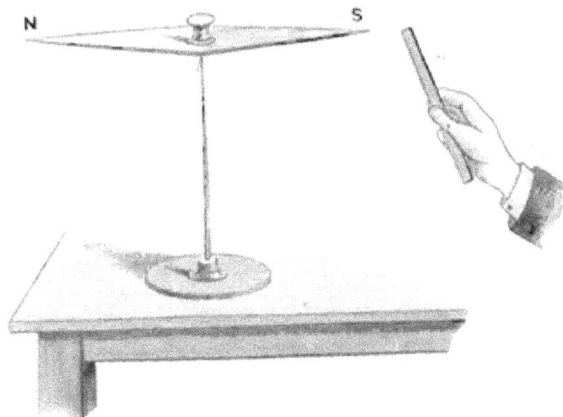

FIG. 11.

with a small brass cap (the bottom part of which is hollow), resting in a horizontal position upon the point of an upright steel rod.

As in previous experiments, it will finally point in a north and south direction.[1]

Take another magnet, and present the poles successively to the north-pointing end of the needle ; one pole of the magnet attracts, and the other pole repels it. Similarly, present the poles to the south-pointing end of the needle ; again, one pole attracts, and the other repels it ; but observe that the pole which *attracts* the north-seeking pole of the needle, *repels* the south-seeking pole.

Distinction between Poles.—From the preceding experiments it appears that there are two opposite magnetic poles, one of which points to the north, and the other to the south. Various names have been given to them, which may be tabulated thus :—

| North-pointing | N-seeking | N | marked | red | + |
| South-pointing | S-seeking | S | unmarked | blue | − |

In this work the terms *north-seeking* and *south-seeking* will generally be used. The reason of this will be given on p. 11. At other times we may speak of the north-seeking pole as the marked pole, and the south-seeking as the unmarked pole.

Exp. 12. Repeat the last experiment, but first suspend the second magnet in a paper or wire stirrup to ascertain the end which points to the north. Mark this with a piece of gummed paper.

Observe that (1) when the marked end is brought near the N-seeking pole of the needle repulsion takes place, *i.e.* two N-seeking poles repel one another.

(2) When the marked end is brought near the S-seeking pole of the needle, attraction takes place, *i.e.* a N-seeking pole and a S-seeking pole attract.

(3) Repeat these experiments with the unmarked end of the magnet.

We, therefore, learn that,

　　　　　like poles repel one another ;
　　　　　unlike poles attract one another.

This is generally called the first law of magnetism.

Exp. 13. This law may be also proved by placing the centre of a magnetic needle over the neutral line of a bar magnet, having

---

[1] Care must be taken that the magnetic needle is removed from the influence of other magnets and of pieces of iron.

previously marked the N-seeking ends. The poles of the needle eventually settle over the opposite poles of the magnet (fig. 12), i.e. attraction takes place between unlike poles, and repulsion between like poles.

Exp. 14, to show the effect of placing the opposite poles of two magnets together :—

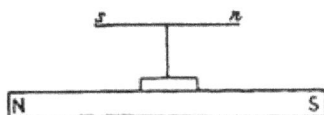

FIG. 12.

Support a key, or any piece of iron or steel, by the bar magnet A, fig. 13. Move a similar magnet, B, along A, so that unlike poles are

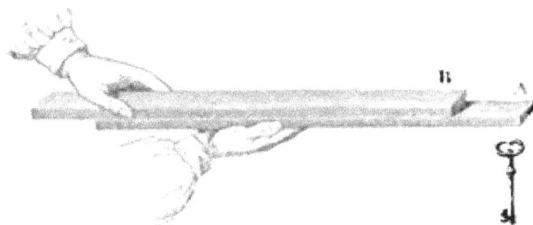

FIG. 13.

in the same direction. When the pole of B approaches the opposite pole of A, the key falls, as though A had lost its magnetism. The two opposite poles therefore neutralise each other. If the two magnets are not equally powerful, or at any rate nearly so, support the key on the less powerful magnet. If the stronger one be used for this purpose, the weaker one is unable to completely neutralise the opposite polarity of A.

Exp. 15, to show the effect of placing like poles of two magnets together :—

Suspend the heaviest possible weight of iron or steel from a bar magnet, move another magnet along it so that similar poles are in the same direction. There is no apparent difference, but it will be found that a greater weight can now be sustained by the combined effect of the two magnets.

The weight which two similar magnets will lift, when their like poles are placed together, is roughly one and a half times that of the lifting power of one.

If a number of magnets, either bar or horse-shoe, be used, having their similar poles close together, they form what is known as a *magnetic battery*. Fig. 14 represents such a battery, in which there are twelve thin magnets—arranged in three sets, each set consisting of four magnets. Their similar poles are bound together by pieces of soft iron, A and B.

FIG. 14.

The Two Poles are Inseparable.—Exp. 16. Heat a piece of watch-spring to a white heat in a blow-pipe flame and immediately plunge it into cold water ; by this means it is made brittle. This process is called *hardening*.

Magnetise the hardened steel by drawing one pole of a magnet from end to end ten or twelve times. Show by means of a magnetic needle that one end has become N-seeking and the other S-seeking.

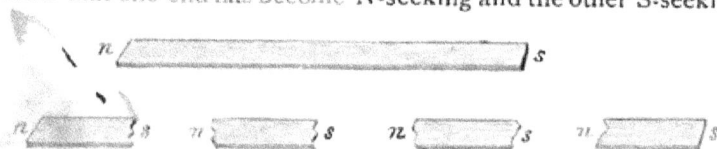

FIG. 15.

Mark the N-seeking end with gummed paper. Break the newly-made magnet into two pieces, and prove that each part is itself a complete magnet, possessing N- and S-seeking poles. If we break the magnet into still smaller pieces, each one will be found to be a perfect magnet, fig. 15.

It is impossible to obtain a magnet with one pole only.

By carrying this principle still further we may imagine a magnet to be composed of an immense number of particles (called molecules), so small that further division by physical means is

FIG. 16.

impossible, having their N-seeking poles all turned one way, and their S-seeking poles the other. If, therefore, the magnet be broken, one face of the fracture must be N-seeking, and the other S-seeking.

Exp. 17. Fill a small test-tube with steel filings. Hold the tube horizontally and draw one pole of an artificial magnet along it a few times, taking care not to shake the tube. Observe that the particles of steel set themselves parallel to the length of the tube, fig. 17. Show that the tube of filings is a magnet by carefully

FIG. 17

approaching a suspended magnetic needle. Shake the tube, and again test ; notice that it is no longer a magnet.

Theories of Magnetism.—(1) The above supposition respecting the arrangement of the particles of a magnet has been called the physical theory of magnetism, or theory of molecular magnets.

(2) Years ago magnetism was explained on the assumption that there were two fluids, mutually attractive and self-repellent. When the fluids were combined they neutralised each other ; when separated the body became magnetised ; in other words, to magnetise a body meant to separate the fluids. We are very uncertain as to the ultimate nature of magnetism ; this theory was, however, useful for the purpose of explaining certain phenomena, although it is now entirely discarded.

(3) Ampère's theory supposes that each molecule of the magnet has a current of electricity

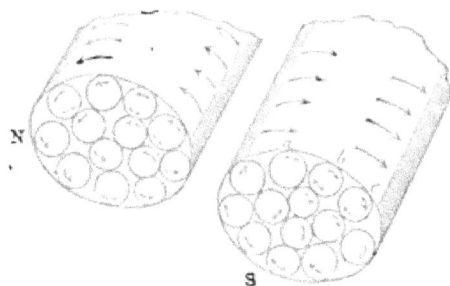

FIG. 18.

circulating round it. The currents are assumed to exist in all magnetic substances (p. 12) ; before magnetisation they move irregularly ; after magnetisation they circulate in parallel directions, and the more perfect the magnetisation the more parallel they become. Looked at from the N-seeking end of a magnet the currents move in an opposite direction to the hands of a watch, and, of course, looked at from the S-seeking end they move in the same direction as the hands of a watch (fig. 18).

This theory will be more fully described in Voltaic Electricity, Chap. XIX.

EXERCISE II.

1. How do magnetic poles act upon each other? What do you understand by a magnetic pole?

2. You are required to demonstrate the law that 'like magnetic poles repel,' and that 'unlike poles attract each other'; how will you do it?

3. Two small bar magnets are so suspended from their ends as to hang parallel to, and at a short distance from each other. Their lower ends, which we suppose to be on the same level, are both north poles. Show by a sketch how they will act upon each other. Show also how they will act when one pole is north and the other south.

4. State the grounds for belief that the smallest particles of a magnet are themselves perfect magnets.

5. How do you figure to yourself the change that occurs in a bar of unmagnetised steel when the pole of a magnet is rubbed along it?

6. A magnet has two poles near its two ends; supposing it to be broken across at the middle, what would be the condition of the two halves?

7. A magnetised steel knitting-needle is broken into three equal parts. Would all three pieces be equally magnetised? If not, how do you suppose they would differ, and why?

8. Can you obtain a magnet with a single pole? Give the experimental grounds of your answer.

9. How would you prove that magnetism is not confined to the poles of a magnet?

10. A piece of steel becomes slightly longer when magnetised. How do you account for this?

Consequent poles.—Although, as previously proved, it is impossible to obtain a magnet with one pole only, it frequently

FIG. 19.

has more than two poles, due to irregular magnetisation. A piece of steel thus magnetised consists, as it were, of several magnets placed end to end, having their poles reversed, as in fig. 19. These *extra poles* are called **consequent poles, consecutive poles,** or **intermediate** poles.

Exp. 18, to produce consequent poles. Take a strip of steel (e.g., a knitting-needle).

(*a*) Draw one pole of a magnet six or eight times from the middle to one end of the needle. (*b*) Draw the *same* pole of the magnet from the middle to the *other end* of the needle. Test the existence of the extra pole by iron filings. The filings are attracted to parts of the strip of steel other than those near the ends.

The earth a magnet.— We have found that a freely-suspended magnetic needle, after oscillating for a time, will finally come to rest so as to point in a direction nearly north and south. To understand this, we must mention the fact that the earth itself is a magnet, acting as though a powerful magnet were placed in it, whose poles are not coincident with, but are situated comparatively near, the true or geographical poles. It therefore *directs* a freely suspended magnet, so that it turns in the direction shown in Exps. 5 and 11.

From the law that 'like poles repel and unlike poles attract,' it will be easily seen that if we speak of the magnetism at the north pole of the earth as north magnetism, that in the north pole of a magnet must be called south magnetism. For this reason we have adopted the terms ' N-seeking ' to denote that pole of a magnet which points northward.

Exp. 19. Allow a **magnetised** knitting-needle, suspended in a paper stirrup as in Exp. 5, to come to rest above a sheet of paper lying on a table. Mark the position of the two ends of the needle. Remove it, and join the points. The line thus drawn points to the magnetic north and south poles. This line is the intersection of the magnetic meridian[1] of the place immediately below the centre of the needle and the earth's surface.

[1] The *magnetic meridian* of any place is the imaginary plane drawn through the zenith (the point in the heavens immediately overhead) and the magnetic north and south points of the horizon. These magnetic north and south points are observed at the place under consideration by the direction of a horizontally suspended magnetic needle when at rest.

Magnetic Substances.—Exp. 20. Plunge a magnet into sawdust, copper or brass filings ; there is no attraction between these substances and the magnet.

We thus learn that some bodies neither attract, nor are attracted by, a magnet.

A body which has the property of attracting, and of being attracted by, a magnet, is called a *magnetic body*.

Besides iron and steel, the other magnetic bodies are cobalt, nickel, chromium, manganese, and cerium. Of the latter, cobalt and nickel are the best, but even they are distinctly inferior to iron or steel in this respect. Many other bodies, e.g., paper, porcelain, &c., are said to be feebly attracted by very powerful magnets.

A magnetic substance which has not been magnetised may be distinguished from a magnet as follows :—

A magnetic substance does not attract unmagnetised iron. It can be attracted by either end of a magnet, but is repelled by neither—in fact, it has no poles.

A magnet has poles, one of which attracts and the other repels one pole of another magnet.

EXERCISE III.

1. Iron and brass filings are mixed together. How can you separate them ?

2. What is the difference between a magnet and a magnetic substance ?

3. I have two pieces of the same steel, of equal size, shape, and hardness ; one piece is a magnet and the other not. State fully the particulars in which they differ from each other.

4. You are to examine, by means of a magnetic needle, the magnetic condition of a piece of soft iron. You are also to examine the condition of a piece of magnetic iron ore of the same size and shape as the iron. State some of the differences which are sure to show themselves between both.

5. Explain what is meant by the term *consequent poles*.

6. A knitting-needle is given you ; how would you magnetise it so that both ends shall be N-seeking poles ? How can you prove that you have succeeded ?

CHAPTER II.

INDUCTION.

Induced Magnetism.—Exp. 21. Bring one pole of a magnet in contact with a piece of soft iron. As we have previously learnt, the magnet, if it be sufficiently strong, will attract and support the iron. Now dip the free end of the iron into iron filings ; they will adhere in tufts (fig. 20), showing that it has become magnetised under the influence of the magnet. Remove the magnet, the filings immediately fall.

FIG. 20.

This influence is exerted even if the iron is not in contact with the magnet. Let us prove this by experiment.

Exp. 22.—Hold the magnet some distance from the iron ; if iron filings be brought in contact with the lower end of the bar a result is obtained similar to that in the last experiment (fig. 21).

The magnetism which thus appears in the piece of soft iron is called *induced magnetism* ; the

FIG. 21.

magnet producing this magnetism is called the *inducing magnet* ; the action is known as *magnetic induction.*

Nature of Induced Polarity.—Exp. 23. Arrange a rod of wrought iron with one end near the north-seeking pole of a magnetised needle (suspended by one of the methods previously described). Bring the N-seeking pole of a magnet near the other end (fig. 22). Observe that the N-seeking pole of the needle is immediately repelled.

The inference from this is that the end of the rod near the needle becomes N-seeking under the inductive action of the magnet, and the other end must therefore be S-seeking, i.e.,

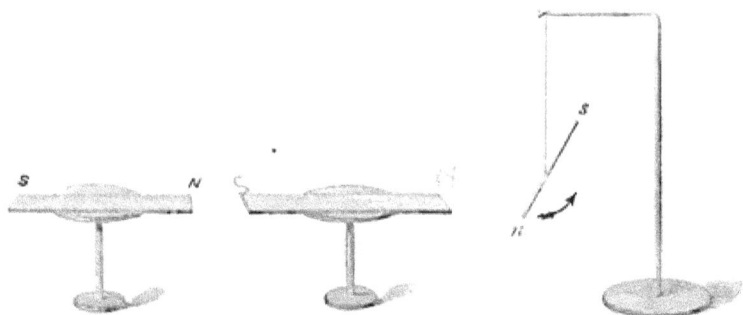

FIG. 22.

the pole of the magnet induces opposite polarity in the end of the rod near to it, while the same polarity is found at the further end.

In this experiment care must be taken that the repulsion of the N-seeking pole of the needle is not directly due to the influence of the magnet itself.

Exp. 24. Suspend a number of needles of equal length, on the N-seeking pole of a strong magnet. Observe how the free ends repel one another.

The nature of the induced polarity can also be shown as follows :—

Exp. 25. Place a horseshoe magnet in a vertical plane, with its poles uppermost. Fasten a silk fibre to a rod of soft iron, so that it hangs horizontally. Suspend the rod over the poles of the magnet, and then bring another magnet near one end of the rod, so that its pole is of opposite name to that of the horseshoe magnet below it. Notice repulsion.

We will now prove that the inductive influence takes place through a number of iron rods.

Exp. 26. Arrange a number of soft iron rods (either in contact or having a space between them), as shown in fig. 23. Bring a pole of a strong magnet near one end of the series. If iron filings be brought in contact with the other end, they will be attracted ;

therefore this piece of iron has become a magnet by induction taking place through the series.

FIG. 23.

Exp. 27. Place a small piece of soft iron in contact with the N-seeking end of a strong magnet. Another piece will be supported if brought in contact with the free end of the first ; this, in turn, will support a third, and so on. The whole series forms what is commonly called a *magnetic chain* (fig. 24). In the figure the N-seeking pole induces S-seeking magnetism in the end of the piece of iron next to it, and N-seeking at the further end ; this induces S-seeking in the near end of the next piece, and so on.

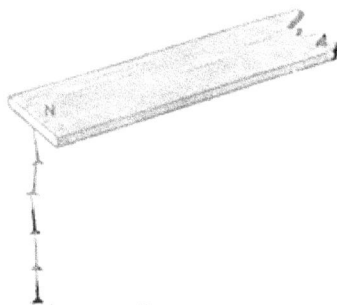

FIG. 24.

Exp. 28. Show, by means of a magnetic needle, that the polarity at the free end of the bottom piece is similar to that of the inducing pole.

Exp. 29, to further illustrate the nature of the induced polarity. Repeat Exp. 24 until no more pieces can be supported. Bring the S-seeking pole of another magnet under the series as in fig. 25. This S-seeking pole will strengthen the polarity of the

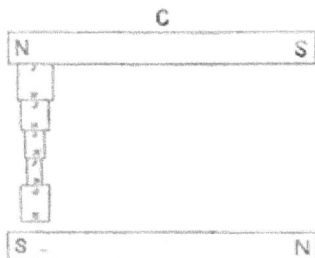

FIG. 25.

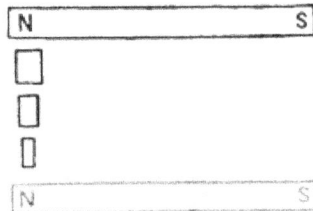

FIG. 26.

pieces if they are suspended from the N-seeking end of the magnet. Test, if this be the case, by adding more pieces. Remove the lower magnet, and the extra pieces fall. If, however, the N-seeking pole be brought under the chain, the pieces at once fall (fig. 26). The reason of this will readily be understood by the student. The

N-seeking pole of the lower magnet induces S-seeking magnetism in the ends of the pieces which were before N-seeking, and *vice versâ*, so that the two opposite magnetisms neutralise each other.

Retention of Magnetism.—Exp. 30. Use pieces of very soft iron to form a magnetic chain. Remove the uppermost piece from the magnet, and observe that the others immediately fall. Test one of the pieces by bringing it near both poles of a magnetic needle. There is no repulsion, therefore all trace of magnetism has disappeared.

These small pieces of soft iron form *temporary magnets*, i.e. they do not retain their magnetism after being removed from the inducing source.

Exp. 31. Repeat the last experiment with small pieces of steel (say steel pens). Holding the top piece, remove the magnet; the other pieces still remain clinging to it; i.e., steel retains its magnetism permanently.

A *permanent magnet* is one which retains its magnetism after being removed from the influence of the inducing magnet.

EXERCISE IV.

1. Why does a magnet attract a piece of iron?

2. How does the middle of a magnet act upon a piece of iron? How do the ends of the magnet act upon the same iron? Does any change occur in the iron when the magnet acts upon it?

3. What is the magnetic condition of a bar of soft iron held near, and parallel to, a bar magnet?

4. Near a ball of perfectly annealed soft iron the north end of a strong steel magnet is placed; what is the action of the magnet upon the ball? What change occurs in the ball when the magnet is withdrawn, and what occurs when the south pole of the magnet, instead of the north, is placed near the ball? Illustrate your answer by diagrams.

5. If a compass needle is deflected when a steel bar is brought near it, how can you find out whether the deflection is due to magnetism already possessed by the bar, or to the bar becoming magnetised by the compass needle at the time of the experiment?

6. Six sewing-needles are hung on one pole of a magnet. Give a drawing of their appearance. Explain.

7. Give an experiment to prove that an induced magnet may become an inducing magnet.

8. What is meant by saying that 'magnets attract magnets only'?

9. A compass-needle and a straight strip of soft iron of the same length as the compass-needle are fastened together so as to be in contact with each other at both ends. Will the force which tends to make the combination point north and south be the same as that which would act on the compass-needle alone? Give reasons for your answer.

10. The **N-seeking pole** of a strong bar magnet is placed some distance from the N-seeking **pole of a** suspended magnetic needle. Repulsion takes place. If, however, it be brought quite close to the needle attraction ensues. Explain this.

11. A bar magnet is held **vertically,** and two equal straight bits **of soft** iron wire hang downwards **from its** lower end. The lower end **of each of** these wires can by itself hold up a small scrap of iron; but if **the lower** ends of both wires touch the same scrap of iron at the same time, **they do** not hold **it up;** what is the reason of this?

12. **Two similar** rods of very soft iron have each of them a long **thread** fastened to one end, by which they hang vertically side by side. On bringing near the iron rods, from below, one pole of a strong bar magnet, **the** rods separate from each other. Explain this.

13. Why is less force required to pull **a small** iron rod away from **the** poles of a powerful **horseshoe** magnet **than** would be required to pull a thick bar of iron away from the poles of the same magnet?

Difficulty of Magnetisation of Steel. —**Exp. 32.** Take a steel knitting-needle and a piece of soft **or wrought iron** (both two **or** three inches long). Dip one **end of each into** iron filings, and **then** bring one pole **of a** magnet **in contact with the** opposite **ends.** Notice that after withdrawal the mass of iron filings attached to the **steel** is smaller than that **attached** to the soft iron. On removing the magnet most of **the** filings still adhere to the steel, **but** they at once drop from the **iron.** They *all* drop from very soft iron.

Steel, **under** induction, **may** be magnetised more easily **and** strongly by smartly tapping **it.** This is probably due to the fact that the molecules are assisted in arranging themselves **in the** direction of their greatest length (p. 9).

Exp. 33. Dip the same end of **the** knitting-needle into iron filings. Bring the same pole of the magnet **in** contact, and while it is under the inductive influence tap it smartly six or eight times. Observe that **a** greater quantity of filings adhere than in the previous experiment, **and on** removing the magnet a still smaller number fall.

The difference between steel and iron in taking up and retaining magnetism is due to the fact that steel possesses a higher *coercive force* or *retentivity* than iron.

C

Coercive force or **retentivity** may be defined as the power which resists magnetisation or demagnetisation.

It is not a force in the ordinary acceptance of the term, so that the term 'retentivity' is preferable to 'coercive force.'

It is extremely feeble in soft iron, but very great in hardened steel.

These results may be tabulated thus :—

—	Under induction	After inducing magnet is removed
Soft iron	is magnetised at once	loses magnetism immediately
Steel	is magnetised more slowly	keeps it permanently

It must be remembered that there is no iron so soft as to have absolutely no retentivity, nor steel so hard as to retain absolutely all its magnetism.

Induction precedes attraction.—In fig. 27 a piece of soft iron is suspended by a thread. As the N-seeking pole of a magnet, N, approaches the iron, induction takes place ; the side of the iron near N has S-seeking magnetism induced in it, and the farther side N-seeking. Attraction takes place between the two opposite poles (N and S), and repulsion between the similar poles ; but as the two opposite poles are nearer than the two similar ones, the force of attraction is greater than the force of repulsion, and therefore the iron is attracted. The exact law respecting the force of attraction or repulsion, and the distance between the two magnets, will be given on p. 22.

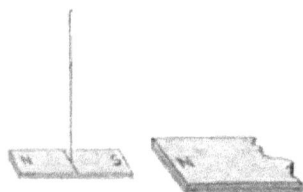

FIG. 27.

Attraction across various Substances.— Exp. 34. Place a large sheet of glass, cardboard, a wooden board, or even the body between a magnet and a horizontally suspended magnetic needle. Observe that the action is exactly similar to that in the previous experiments, where no body (except air) was interposed.

The same result may be shown as follows :—

Exp. 35. Magnetise a knitting-needle (two or three inches long), and suspend it by a fibre of raw silk, so that it hangs horizontally. If moved from its position of rest it will make a certain number of oscillations in a fixed time, say one minute. If, however, the N-seeking pole of a magnet be brought towards the S-seeking pole of the needle, after the latter has been moved from rest, it will make a greater number of oscillations than before. Count the number of oscillations made in one minute, when each of the substances mentioned in the last experiment is interposed between the needle and the magnet. The distance between the two magnets remaining constant in each case, the number of oscillations will be equal.

Exp. 36. Repeat the last experiment with a large sheet of soft iron interposed. The number of oscillations made by the needle in the same time is less than in that experiment; in fact, it approaches that obtained when the needle oscillated under the earth's influence alone. If the iron was quite soft and very thick no action would take place across it.

It, therefore, appears that magnetic force acts across all bodies with the exception of iron and the other magnetic substances. Again, the body, unless it is a magnetic substance, across which the action is transmitted, is not affected by the magnetic force, e.g., a person feels no sensation when it passes through his body.

Keepers and Armatures.—In order to preserve the power of magnets when not in use, they are provided with pieces of soft iron—called *keepers* or *armatures*—which are placed over the poles.

Fig. 28 represents the arrangement with two bar magnets. The magnets themselves are arranged parallel to each other, the N-seeking pole of one being near to, but not touching, the S-seeking pole of the other. The keepers, A B, are then placed in contact with the poles of the magnets.

Fig. 28.

It is preferable to arrange the magnets in a box, a thin wooden rod being placed between them.

The reason of this preservative power will be easily understood. During contact each pole of the magnet induces oppo-

site polarity in that part of the keeper nearest to it. Consider
the N-seeking pole of one magnet. It is acted on by the
induced S-seeking pole of the keeper, as well as by the
S-seeking pole of the other magnet. The opposite polarities,
therefore, attract each other, and thus tend to preserve that
arrangement of the particles of the magnet in which their
N-seeking poles lie towards one direction, and their S-seeking
poles towards the other.

Fig. 29 represents a horseshoe magnet with the keeper
attached. The principle of the preservative power is similar
to that of the bar·magnets described above.

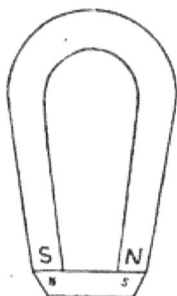
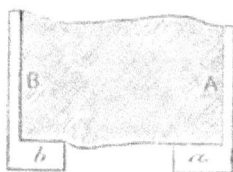

FIG. 29.   FIG. 30.   FIG. 31.

Figs. 30 and 31 represent the arrangement for a natural
magnet.

Having ascertained the position of the poles by means of
iron filings, the lodestone is ground so that the two faces
which contain the poles A B are parallel. They are then fitted
with two plates of soft iron provided with projecting feet, *a b*.
These form the armatures of the lodestone. They are bound
together by brass caps (fig. 31), one at the top and the other
just above the feet, *a b*. The keeper *a' b'* is then added.

Exercise V.

1. What is the difference between iron and steel as regards their acceptance and their retention of the magnetic condition? How is their difference usually explained?

2. You have two similar rods, one of steel and the other of iron ; you have also a bar magnet and some iron nails. Describe exactly some experiment which would enable you to distinguish the steel rod from the iron.

3. The pole of a magnet is brought within an inch of one side of a sphere of very hard steel, suspended from a string. It manifestly attracts the steel, but it is not quite able to draw it into contact. A sphere of iron of the same weight is now substituted for the sphere of steel, and the magnet is found able to draw this new sphere quite up against itself. Explain this difference of action.

4. Two bars of soft iron are so placed to the east and west of the N-seeking pole of a compass-needle that the needle still points north and south. If the iron to the east of the needle be replaced by a bar of hard steel of exactly the same size and shape as itself, will the direction in which the magnet points be altered? If so, in which direction will it move, and why?

5. If you buy a horseshoe magnet you generally find a piece of iron fitted to the two ends. What is its use?

6. What would you do to preserve the magnetism of two bar magnets when not in use? Explain.

7. What is the best arrangement for three bar magnets when not in use?

CHAPTER III.

LAWS OF MAGNETIC FORCE. MAGNETIC FIGURES.

Coulomb's Torsion Balance.—By aid of the torsion balance Coulomb proved that *the force of magnetic attraction or repulsion varies inversely as the square of the distance.*

The construction of this apparatus will be understood from fig. 32. It consists of a glass case, having two apertures in the top (1) near the edge, to admit a magnet A, (2) at the centre, into which a narrow glass tube is fitted, provided with a brass cap. This cap is shown in the side figure, and consists of two discs — one, D, fixed to the tube, and having its circumference divided into 360°; the other, E, movable about its axis. It has a mark, c, by means of which the number of degrees can be read, through which it has been turned from the zero on D; and is provided with two uprights connected by a cross-piece, to which a fine silver wire is attached, which carries the small magnetic needle a b. On the side of the case there is a graduated scale, which shows the angle through which the needle a b turns.

The principle involved in the use of this instrument

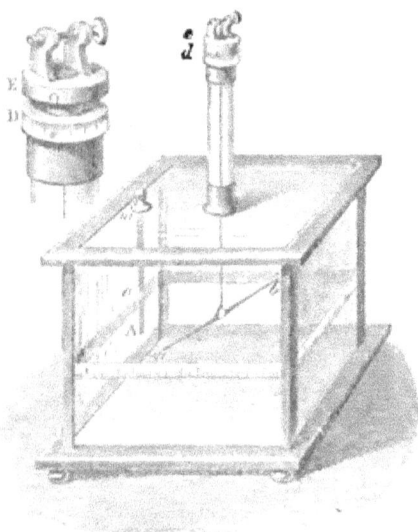

Fig. 32.

depends upon the well-known law that the force of torsion (the force with which the wire is twisted) is proportional to the angle of torsion.

In a particular experiment Coulomb first adjusted the apparatus so that the magnetic needle was in the magnetic meridian without the wire being twisted. This was done by putting in the needle and roughly adjusting, then replacing it by a copper needle of equal weight and again adjusting. The operations were repeated until the needle lay exactly in the magnetic meridian without twisting the wire. He next ascertained the number of degrees the disc *e* must be turned to make the needle move 1° from the meridian. This he found to be 35°.

The magnet A was then introduced so that its lower pole repelled that of the needle. The needle was repelled by this means through 24°. This force = 24° × 35° of the torsion head ; the wire also has an actual twist of 24°. Thus the total force is proportional to 24° × 35° + 24° = 864°. Lastly the disc *e* was turned so as to bring the needle *a b* to half the distance, i.e. the two poles were now 12° apart. This required eight complete revolutions of the disc, i.e. a twist of 8 × 360° = 2880°, but the bottom of the wire was still twisted 12° more than the top, therefore the force producing this twist = 12° × 35° = 420°, and there is an actual twist of 12°, so that the total force = 2880° + 420° + 12° = 3312°.

Now 3312 is nearly four times 864 ; therefore at half the distance the repulsive force is four times as great. If the distance had been $\frac{1}{3}$, the force would be nine times as great.

Tabulating these results we have :—

Distance	1	2	3	4	etc.
Force of repulsion	1	$\frac{1}{2^2}$ $=\frac{1}{4}$	$\frac{1}{3^2}$ $=\frac{1}{9}$	$\frac{1}{4^2}$ $=\frac{1}{16}$	etc. etc.

Strength of a Magnet.—The strength of a magnet is the amount of free magnetism in either pole, i.e. the amount of magnetism not neutralised by opposite magnetism in adjacent particles of the body. It is measured by the magnetic force exerted upon other magnets. The method will be understood by means of an example, carefully remembering that when a body oscillates so as to "perform each oscillation in the same time, *the force producing the oscillation is proportional to the square of the number of oscillations in a given time.* In a particular experiment a small magnetic needle (say three-quarters of an inch long) was freely suspended, which when slightly disturbed from its position of rest made nine oscillations in one minute. These oscillations are due to the action of the earth's magnetism, which is therefore measured by the number 9^2 or 81.

A magnet, say 16 inches long, which we will call A, was then placed in the meridian with its S-seeking pole two inches from the N-seeking pole of the needle (fig. 33), which then oscillated thirty times in one minute.

Fig. 33.

Afterwards another magnet, B, was placed similarly, and the needle then oscillated twenty times in one minute. Whence

the strength of A : strength of B:: $30^2 - 9^2$: $20^2 - 9^2$

$$:: 900 - 81 : 400 - 81$$
$$:: 819 \qquad : 319$$
$$:: \frac{819}{319} \qquad : 1$$

Therefore the strength of A is nearly $2\frac{1}{2}$ times that of B.

This method of finding the comparative strengths of two magnets is called the *method of oscillations.*

Laws of Magnetic Force.—(1) Like magnetic poles repel one another, unlike magnetic poles attract.

(2) The force between two magnetic poles is directly proportional to the product of their strengths,[1] and is inversely proportional to the square of the distance between them.

[1] A magnetic pole is said to be of unit strength when, placed at a dis-

This last law may be expressed in equational form thus :—

$$f = \frac{m \times m'}{d^2}$$

where m and m' = the respective strengths of the poles
and d = distance between them.

Lifting Power.—The lifting power of a magnet must not be confounded with its strength. It depends upon (1) the form of the magnet, (2) its strength.

Horseshoe magnets lift a greater weight than bar magnets of the same size and strength ; this arises from the fact that both poles act upon the weight.

The lifting power is increased in a peculiar manner by gradually adding to the weight borne by the magnet. If, however, the weight be suddenly removed, the magnet at once loses this extra lifting power.

EXERCISE VI.

1. The force which one pole of a magnet exerts upon a pole of another magnet increases as you decrease the distance between them. Give the exact law and its experimental proof.

2. How would you prove that two long bar magnets are of equal strength ?

3. I magnetise a piece of watch-spring. How would you arrange it so as to obtain the greatest lifting-power ?

4. A pole of one magnet has a strength of 20 units, and a similar pole of another magnet has a strength of 5 units. They are placed 5 centimetres apart. What is the force of repulsion between them ? *Answer:* 4 dynes.

5. A pole of a magnet has a strength represented by 45 units, and acts with a force of 30 dynes upon another pole 6 centimetres distant. What is the strength of that pole ? *Answer:* 24 units.

Lines of Force.—The space through which the influence of a magnet extends is called the **magnetic field** of that **magnet.** The magnetic force increases as we approach the poles, and diminishes as we recede from them. Indeed at every point it has a definite intensity depending upon the distance from the

tance of one centimetre from a similar pole of equal strength, it is repelled with a force of one dyne. (For an explanation of the term *dyne* see page 97.)

poles ; and it has a well-defined direction at every point, as indicated by what is called **the line of force** passing through the point.[1]

The general distribution of these lines of force may be experimentally exhibited in the following way :—

Exp. 37. Place a sheet of cardboard above a magnet resting upon a table. Sprinkle iron filings from a pepper castor or a muslin bag over the cardboard. As the filings fall, gently tap the cardboard, and notice that they arrange themselves along certain curves (fig. 34).

These curves represent *the lines of force*, or, as they are more correctly called, *the lines of induction*. By induction each particle of iron becomes a magnet ; its N-seeking pole being towards the S-seeking pole of the magnet, and *vice versâ*.

The following diagrams show the lines of force in various cases as indicated by the arrangement of iron filings. Observe how the lines form curves from pole to pole. When two bar magnets are used, their appearance, of course, depends upon the manner in which the poles are placed with regard to each other. Obtain the curves experimentally as explained in Exp. 37.

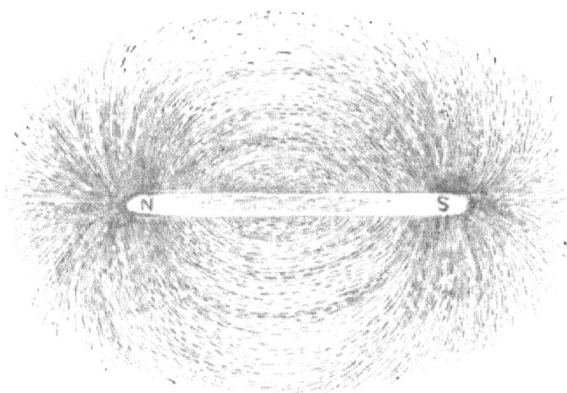

FIG. 34.—ARRANGEMENT ABOUT A SINGLE BAR MAGNET.

[1] The actual direction of the magnetic force at any point is in the line which touches 'the line of force' at that point.

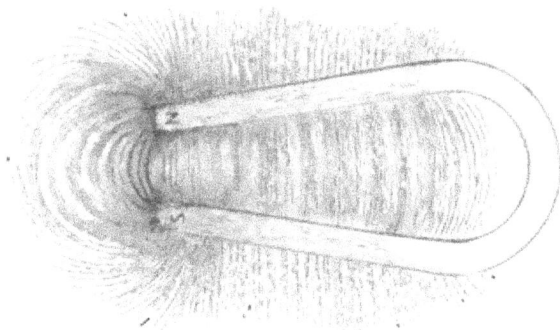

FIG. 35.—ARRANGEMENT ABOUT A HORSESHOE MAGNET.

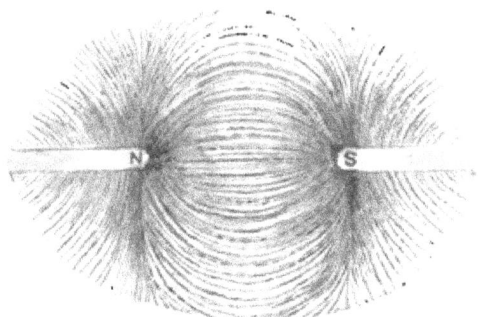

FIG. 36.—ARRANGEMENT ABOUT THE *two Dissimilar Poles* OF TWO BAR MAGNETS.

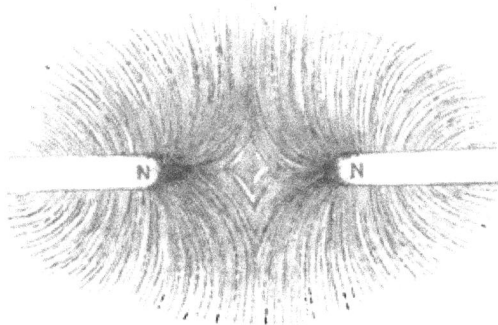

FIG. 37.—ARRANGEMENT ABOUT THE *two Similar Poles* OF TWO BAR MAGNETS.

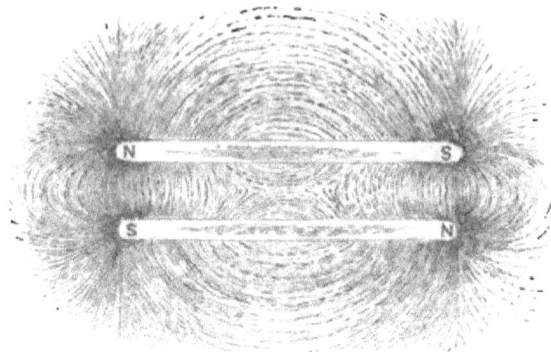

FIG. 38.—ARRANGEMENT ABOUT TWO PARALLEL BAR MAGNETS WITH THEIR *Dissimilar Poles* ADJACENT.

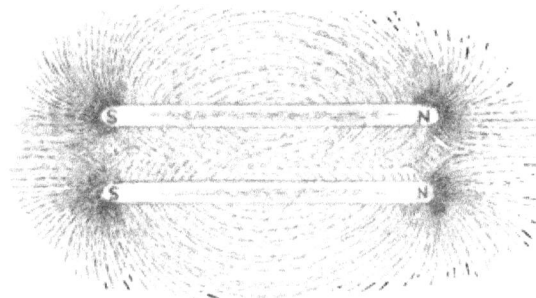

39.—ARRANGEMENT ABOUT TWO PARALLEL BAR MAGNETS WITH THEIR *Similar Poles* ADJACENT.

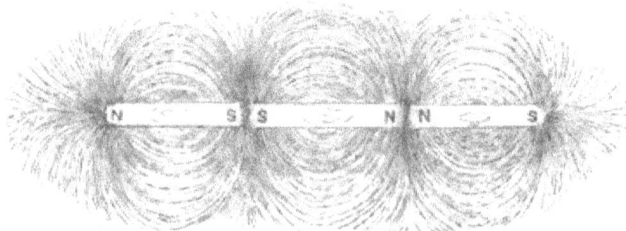

FIG. 40.—ARRANGEMENT ABOUT *Consequent Poles* OF A BAR MAGNET.

FIG. 41.—ARRANGEMENT ABOUT *One Pole* OF A BAR MAGNET.

A permanent record of these *magnetic figures* can be made in various ways.

(1) Carefully pour a solution of potassium cyanide over the cardboard or paper ; the potassium cyanide combines chemically with the iron, forming a precipitate of Prussian blue.

(2) Soak paper with a solution of nut-galls. Carefully place the paper over, and in contact with, the filings. By the chemical action of the iron, black ink is produced, which marks their position.

(3) The best method is to previously gum and dry the paper. After the filings have settled, gently blow a jet of steam over the paper. The steam moistens the gum, which is then allowed to harden, thus fixing the filings in their places.

It is advisable to place the paper on a sheet of glass.

Exp. 38, to show the position a magnet takes in the field of another magnet.

Having obtained a map of the curves produced by a bar magnet, take a small magnetic needle and place it across one of the lines of force. Observe that it immediately sets itself in the direction of the line of force on which its centre lies.

EXERCISE VII.

1. What is meant by a magnetic field, and what is meant by a magnetic line of force ?

2. Sketch a bar magnet and letter its poles. Show by dotted lines the manner in which iron filings arrange themselves round the bar magnet. Do the same with a horseshoe magnet.

3. Two bar magnets are placed upon a table parallel to each other, and with their north poles turned in the same direction. Over the magnets is placed a sheet of glass, and over this again a sheet of smooth paper. From a little sieve you carefully scatter iron filings over this paper. Show by one sketch the manner in which the filings will arrange themselves, and show by another sketch the change that occurs in the arrangement of the filings when one of the magnets is reversed.

CHAPTER IV.

METHODS OF MAGNETISATION.

MAGNETS may be made from bars of steel by several methods :

1. by rubbing them with permanent magnets according to certain methods to be described,
2. by using electro-magnets instead of permanent magnets,
3. by passing electric currents round them,
4. by the inductive action of the earth.

There are three methods of magnetisation by the action of permanent magnets, which are known as :—

(*a*) Single touch.
(*b*) Separate or divided touch.
(*c*) Double touch.

Single Touch.—Exp. 39. Place a steel knitting-needle or a small steel bar upon a table. Draw one pole of a magnet from one end of the bar to the other, taking care to move it always in the same direction. In fig. 42 the arrows indicate the movement of the magnetising magnet. Turn the bar over after rubbing one side ten of twelve times, and treat the other side similarly. The pole at the end of the bar where the magnet leaves it is always of the opposite name to the magnetising pole.

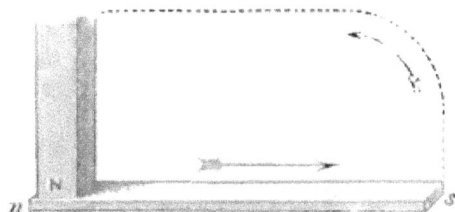

FIG. 42.

Thus, if a N-seeking pole of a magnet be used, the end where the magnet leaves the bar becomes S-seeking, that where it is first placed N-seeking.

Take the newly made magnet, and prove that these statements are true by means of a magnetic needle.

This method of single touch is suitable only for magnetising small bars, such as compass-needles. Unless great care is taken there is a tendency to develop consequent poles (see p. 10).

Separate or divided Touch. — Exp. 40. Place a steel bar so that its ends rest on the opposite poles of two bar magnets. Place the opposite poles of two other magnets at the middle of the bar at an inclination to the bar of about 20°, taking care that the poles are similar to those of the magnets below (fig. 43). Draw them

FIG. 43.

simultaneously from the middle of the bar to the ends. Lift them and place them again at the middle. Repeat this operation ten or twelve times. Turn the bar over, and treat the other side similarly.

This method probably produces the most regular magnets.

Double Touch. — Exp. 41. As in the last experiment place the bar to be magnetised so that its ends rest on the opposite poles of two bar magnets. Fasten a small piece of wood or cork between the magnetising magnets (fig. 44) so as to keep their opposite poles

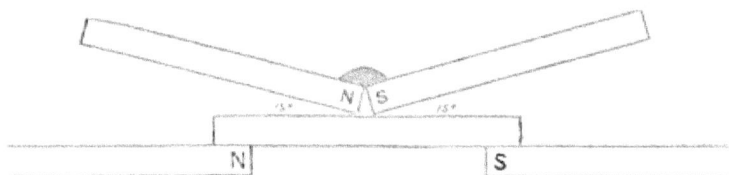

FIG. 44.

at the same distance from each other. The inclination of these magnets to the bar should be about 15°.

Move the magnets to one end of the bar, then back to the other (not lifting them). Repeat the process ten or twelve times, taking care to leave off at the middle of the bar, so that each half is rubbed an equal number of times. In this method, too, the bar should be turned over, and the process repeated.

This method makes the most powerful magnets; its disadvantage, however, is that it has a tendency to produce consequent poles.

Magnetism by the Electric Current.—The method will be merely indicated here, a more complete treatment being given after some knowledge has been gained of current electricity.

Exp. 42. Wind copper wire in the form of a spiral round a glass tube, as in fig. 45. Place a bar of steel inside the tube.

FIG. 45.

Pass a current of electricity through the spiral by fastening the ends to the two terminals of a voltaic battery. Observe that the bar becomes magnetised, by means of iron filings.

Exp. 43. The following method is frequently adopted practically. Bend silk-covered copper wire into a coil (fig. 46). Attach the ends to the terminals of a fairly strong voltaic battery. Move the coil from one end to the other of a bar (or horseshoe-shaped piece) of steel, taking care to move it always in the same direction.

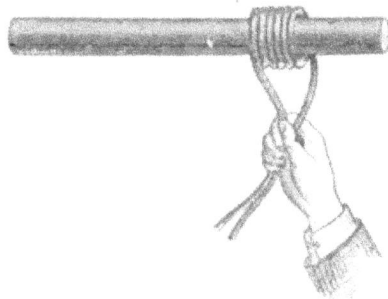

FIG. 46.

Magnetisation by Electro-Magnets.—An electro-magnet is simply a bar of soft iron, generally in the shape of a horseshoe, round which a coil of insulated copper wire is wound, through which a current of electricity can be passed.

They are often used to magnetise bars of steel, as they are much more powerful than ordinary permanent magnets.

Exp. 44. (1) Move a steel bar from one end to the other along one pole of an electro-magnet, round which a current is flowing (fig. 47). (2) Move it in the opposite direction along the other pole. If the bar be drawn from end to end across the N-seeking pole of an electro-

FIG. 47.

D

magnet, the end of the bar which leaves that pole becomes S-seeking. This S-seeking pole is then placed on the S-seeking pole of the electro-magnet, and drawn across it, making the other end N-seeking, thus further developing its magnetism.

It is a common practice of magnet-makers to fix the electro-magnet into a board. The student will understand that a steel bar can be magnetised by an ordinary horseshoe magnet in a method exactly similar to the one just described.

Magnetism by the Earth's Induction.—Exp. 45. (1) Draw a horizontal line A B on a sheet of cardboard ; make an angle A O C of $67\frac{1}{2}°$ with the line A B (fig. 48).[1]

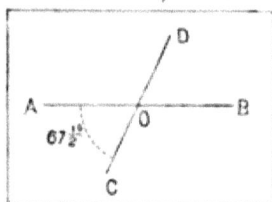
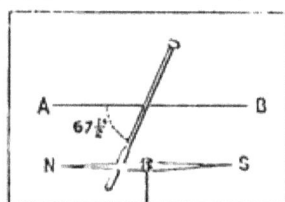

Fig. 48.   Fig. 49.

(2) Allow a magnetic needle to become steady.

(3) Fix the cardboard parallel to the vertical plane which passes through the magnetic axis of the needle (fig. 49).

(4) Place a poker on the line C D.

(5) Strike the poker smartly a few times while it occupies this position.

(6) Present the lower end of the poker to the N-seeking end of a magnetic needle. Notice repulsion. We therefore learn that the poker has acquired magnetic properties, and that the downward end becomes N-seeking.

If a soft iron bar be used it should be tested by bringing the N-seeking pole of a magnetic needle near the lower end while it occupies this position. If the bar be of steel or cast iron it will require a little time to become thoroughly magnetised. The reason will be understood from previous explanations.

[1] This is the value of the angle at London in 1889.

A poker will, however, become magnetised, even if it be held perpendicularly ; the lower end becomes N-seeking.

Magnetism after Heating.—Exp. 46. Place a piece of steel, heated to redness, between the opposite poles of two magnets, or between the poles of an electro-magnet. Allow it to cool, and prove that it has become magnetised.

The magnetic condition may be acquired by a red-hot bar of steel if cooled whilst lying in the magnetic meridian.

Magnetic Saturation.—Many of the methods above described will produce more magnetism in a bar of steel than it can retain permanently, i.e. it will become *supersaturated*, although the excess will very quickly disappear. A magnet is said to be *saturated* when it is magnetised to a strength which it can permanently retain.

Soft iron after magnetisation retains a very small quantity of magnetism, which is known as *residual magnetism*.

Loss of Magnetism.— Magnetism may be destroyed or weakened in several ways :—

(1) by placing magnets when not in use with the similar poles near together. In this case the S-seeking magnetism of one magnet induces N-seeking in the other ;

(2) the magnetism óf the earth may induce opposite polarity ;

(3) by rough usage, such as a heavy blow or fall. This disturbs the arrangement of the molecules ;

(4) by heating. If a magnet be made red-hot its magnetism is destroyed.

Effects of Magnetism.—(1) It is found that when a bar of steel or iron is magnetised, it becomes very slightly longer, and, as its volume remains the same, it therefore becomes very slightly narrower. This is due to the fact that the particles of the bar set themselves in their longest directions during and after magnetisation.

(2) A faint click is heard at the moment of magnetisation and demagnetisation of a bar.

(3) Heat is produced when a bar is rapidly magnetised and demagnetised.

Diamagnetism. — Faraday, in 1845, discovered that all bodies are acted on by powerful magnets—some being attracted, others repelled. Those that are attracted he called ***paramagnetic***, or merely *magnetic*; those that are repelled he called *diamagnetic*. In his experiments he used a powerful electromagnet, suspending various substances between the poles. A small rod of iron, for example, set itself *axially*, i.e., in a line joining the two poles. The same result took place with other magnetic bodies. He observed, however, that certain substances were repelled by the poles of the magnet into a position at right angles to this direction, e.g., *equatorially*. These were the diamagnetic bodies, of which the principal are bismuth, phosphorus, antimony, glass, sulphur, water, zinc, tin, copper, lead, gold.

EXERCISE VIII.

1. Describe how you would impart the magnetic power of the lodestone to a piece of steel.

2. A steel sewing-needle is drawn over the north-seeking pole of a magnet from eye to point; what is the subsequent condition of the needle? The point is presented to the north end of a mariner's compass-needle; what occurs?

3. You have a bar magnet and a steel knitting-needle, one end of which has been marked by having been dipped in ink. Say exactly what you would do in order to magnetise the knitting-needle, so as to make the marked end a north-seeking pole, and the other end a south-seeking pole; and how could you find out whether you had succeeded?

4. You possess a small magnetic needle, with the end which points to the north marked N., and that which points to the south marked S. An iron poker which has been perfectly annealed is placed upright, and the magnetic needle is brought first near the bottom of the poker, and then gradually raised past the centre to the top. The poker acts upon the needle, and you are required to describe its action.

5. How do you account for the fact that chisels and other tools are often found to be magnetised?

6. A steel knitting-needle held vertically and struck with a hammer becomes a magnet. Why is this? How does the striking and position affect it? You bring the lower end near the north-seeking pole of a compass-needle. What happens?

7. Perform some experiment to show the effect of heat upon a magnet.

8. Mention the two classes into which bodies are divided with respect to their attraction or repulsion by the poles of a magnet. Give a list of the principal substances in each class.

CHAPTER V.

TERRESTRIAL MAGNETISM.

The earth a magnet.—The student has already learnt that the earth is itself a magnet. **It behaves as** though **a powerful** bar magnet, whose poles **are** situated comparatively near **the** geographical north and south poles, lay within its **mass.** The magnetic **north pole** was found by Sir James Ross to **be situated** in Boothia Felix, 96° 46′ west longitude, and 70° 5′ north latitude. The **south** magnetic **pole has not yet been reached, but** it is situated about 166° east longitude and 76° south latitude. We can make **an** approximate representation of the magnetic condition of **the** earth **by** placing **a bar** magnet **within a** wooden globe, **in** such a manner that the centre of the magnet coincides with the centre of the globe, its S-seeking pole being $17\frac{1}{2}°$ to the west of the point which marks the geographical north pole. The reason of the S-seeking pole being near the geographical north pole has been explained on p. 11. In fig. 50 let N be the **true north ;** then the position of the magnet is represented as in the diagram. E Q is the geographical equator and E′ Q′ the magnetic equator. The polarity of the **northern** hemisphere is, however, **called** northern or boreal, **while** that in **the southern** hemisphere is known as southern or austral polarity.

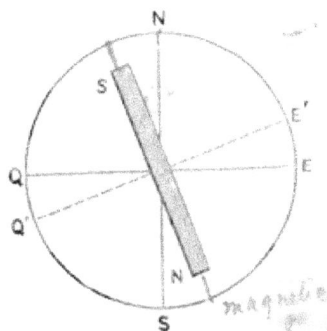

FIG. 50.

Magnetic elements.—To know fully the terrestrial magnetism at any place we must possess **a** knowledge of—

    (1) The declination ⎫
    (2) The inclination ⎬ at any place.
    (3) The intensity ⎭

These are known as the *terrestrial magnetic elements,* or more commonly the *magnetic elements* of the place.

. **Declination or Variation.**—**Exp. 47.** Repeat Exp. 19. (1) Mark the end of the line under the N-seeking end of the needle N, and that under the S-seeking pole, S, and let O be the middle point (fig. 51).

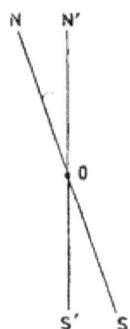

(2) At the point O in the line O N, and on the right-hand side of it make an angle N O N′ of about $17\frac{1}{2}°$.[1]

(3) Produce N′O to S′.

The line S′ N′ points to the true or geographical N and S poles. The angle N O N′ is the **declination** of the place O.

From this experiment we have, therefore, learnt two facts :—

Fig. 51.

(1) That the declination of a place is the angle between the magnetic meridian and the geographical meridian.[2]

(2) How to draw a line true N and S, knowing the declination at the place.

The declination varies at different places on the earth's surface. At present it is west in Europe, but east in many places in Asia and in North and South America. In some places a needle points exactly to the geographical north pole—i.e., the geographical meridian coincides with the magnetic meridian.

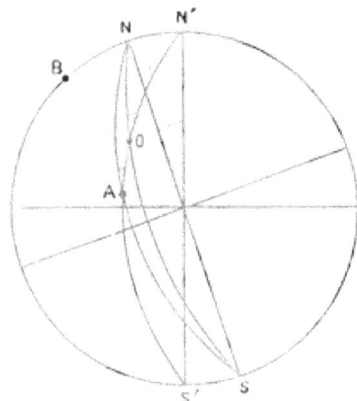

Fig. 52 will show roughly why the declination varies at different places. In practice, however, the variation at different places does not quite follow the theoretical

Fig. 52.

---

[1] The value of this angle varies at different places ; $17\frac{1}{2}°$ is the value at London in 1889.

[2] The geographical meridian of any place is the plane passing through the N and S poles and through the zenith of the place.

consideration here given; in fact, it must be determined by actual observation. For an account of the method of making these observations the student must consult a more advanced work. The diagram will assist the student in grasping the reason of the changes in declination.

Let N'OS' be the geographical meridian of places O and A.
  „ NOS   „   magnetic   „   „ the place O.
  „ NAS   „   „   „   „   „ A.

Then ∠ NON' is the declination at O.
And ∠ NAN' „   „   „ A.
But NON' is the exterior angle of the triangle NAO, and therefore NON' is greater than NAN'.

At a place B situated both in the geographical and magnetic meridians there will be no declination.

Not only does the variation alter at *different places*, but it alters at the *same place*.

These variations are of two kinds :—

(*a*) Regular, including secular, annual, and diurnal.
(*b*) Irregular or accidental.

• **Secular Variations.**—At any particular place the direction of the compass-needle undergoes a gradual change ; the needle at one time pointing to the west of true north, and at another time to the east. This change of declination extends over centuries, during which period it passes from its maximum westward position to its maximum eastward position. We have no record of the variations previous to the year 1580. In that year the N-seeking pole of the needle at London pointed nearly 11° east of the geographical north. The declination then gradually decreased, until in 1657 there was no declination. The N-seeking pole then began to move to the west, attaining its greatest westerly declination (24° 30') in 1816. At the present time (1889) the declination is slowly decreasing.

From the following table a more complete notion of the secular variations at London will be obtained :—

Year	Declination	Year	Declination
1580	11° 17′ E.	1816	24° 30′ W.
1634	4° 0′ E.	1868	20° 33′ W.
1657	0° 0′	1882	18° 22′ W.
1705	9° 0′ W.	1888	17° 40′ W.
1760	19° 30′ W.		

The mean annual change of declination is nearly 7′.

Annual Variations.—The needle is also subject to small annual variations (about 15′ to 18′). In London this variation is greatest at the vernal equinox ; it then diminishes, becoming least at the summer solstice, after which it again increases during the next nine months.

Diurnal Variations.—In very sensitive instruments the needle is observed to have a daily motion. In England the N-seeking pole of a needle moves westward every day, from 7 A.M. to about 1 P.M. It then begins to move eastward, and continues to move in this direction until about 10 P.M. It approximately retains this position until sunrise.

Isogonic and Agonic Lines.— Charts have been prepared in which the places having equal declination are joined by a line. Such lines are called *Lines of equal declination*, or **Isogonic Lines.** Similarly, the line joining places where there is no declination is called the **Agonic Line** (see Map, p. 41).

Irregular Variation. — It sometimes happens that the needle is suddenly disturbed by several causes—e.g., volcanic eruptions, earthquakes, and the aurora borealis. These irregular and accidental disturbances often affect the needle over a considerable area, and are known as **magnetic storms.**

Inclination or Dip.—Exp. 48. Fasten a thread of untwisted silk to the middle of an ordinary knitting-needle, by means of hot shellac. Cause the needle to hang horizontally, filing one end if necessary. Magnetise it, and observe that it sets itself in the magnetic meridian, and moreover that its N-seeking pole dips downwards.

This phenomenon may also be observed by means of the apparatus shown in fig. 53, which consists of a magnetic needle suspended at its centre of gravity so as to move freely in a

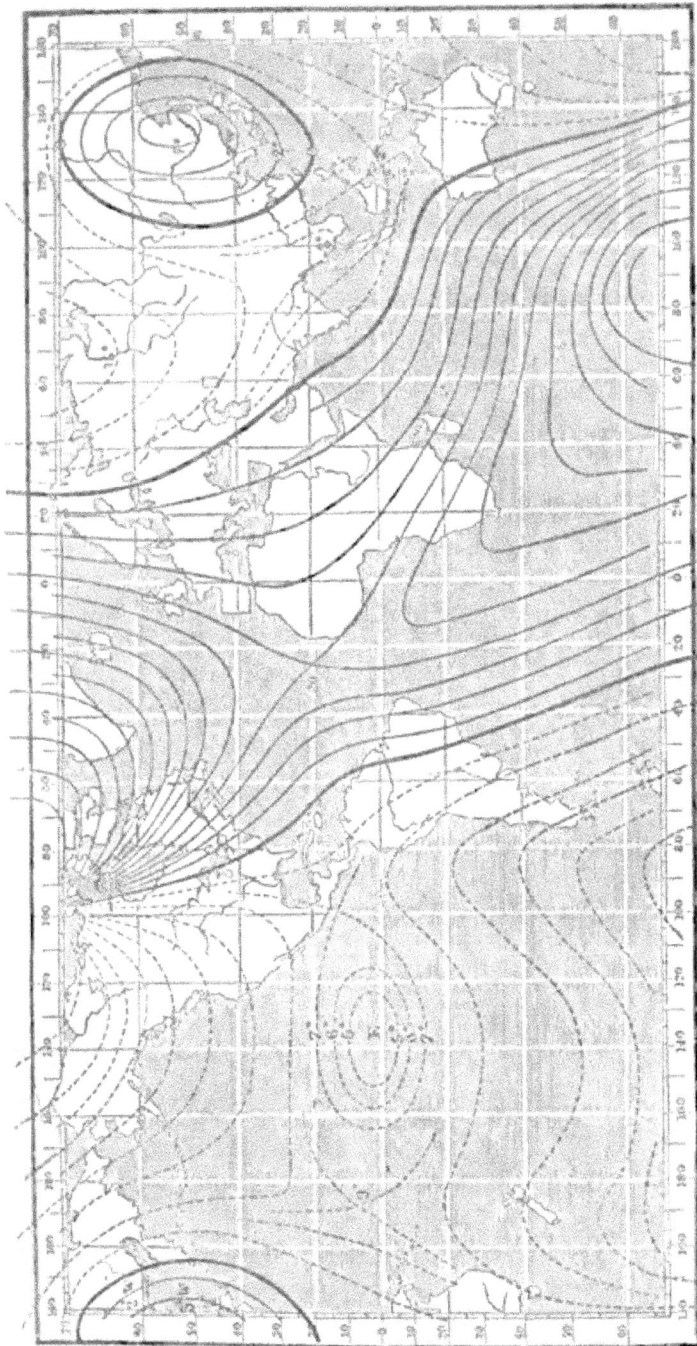

LINES OF EQUAL MAGNETIC DECLINATION

vertical plane about a horizontal axis. Such an arrangement is called a **dipping-needle**, and is generally provided with a quadrant, graduated in degrees.

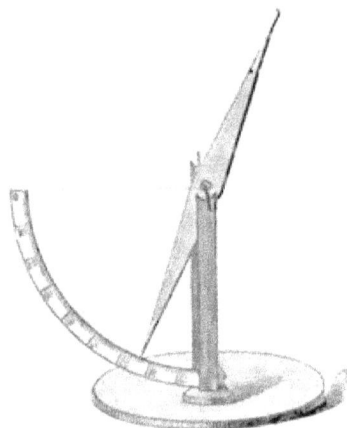

FIG. 53.

Exp. 49. Place the instrument so that the vertical plane in which the needle moves is in the magnetic meridian. Allow the needle to become steady. Read the angle it makes with the horizon, i.e. the angle between it and the horizontal line drawn through the point of suspension.

This angle is called the *dip* or *inclination* of the needle. Dip may, therefore, be defined as the *angle which a magnetic needle, suspended so as to move about a horizonal axis, makes with the horizon when the vertical plane in which it moves coincides with the magnetic meridian.*

The Dipping-Needle or Inclination Compass.—A more delicate form of the instrument is shown in fig. 54. It consists of

(*a*) a graduated horizontal brass circle *m*, supported on three legs, provided with levelling screws.

(*b*) Above this is a plate, A, moving about a vertical axis, which supports

(*c*) a vertical graduated circle, M, upon which the inclination is measured. This moves with the plate A, and is capable of horizontal rotation.

(*d*) A vertical magnetic needle supported on a horizontal axis in the centre of the graduated circle M.

(*e*) A spirit-level, *n*, fixed to the plate A.

To ascertain the dip the instrument is

(1) made level by turning the screws until the air-bubble is in the centre of the spirit-level.

(2) The plate A is then turned on the circle *m* until the needle is vertical. (The plane of the needle is then at right angles to the magnetic meridian.)

(3) A is then turned through 90° on the circle *m*. (The needle is now brought into the magnetic meridian.)

FIG. 54.

(4) Observe the angle between the point *a* of the needle and the zero point (near *d*) on the circle M.

Exp. 50. (*a*) Place the dipping-needle in any plane other than that of the magnetic meridian. Observe that the dip increases.

(*b*) Place it at right angles to the magnetic meridian. Notice that the needle is vertical, i.e. the dip is 90°.

Dip, like declination, varies in different places on the earth's surface. At the magnetic poles the needle is vertical, i.e. the dip is 90°. The dip in London at present (1889) is 67° 25′ nearly.

Isoclinic and **Aclinic Lines.**—The lines which join places at which the dip is the same are called **isoclinic** lines. The line which joins places at which there is no dip is called the **magnetic** equator, or aclinic line (see Map, p. 44).

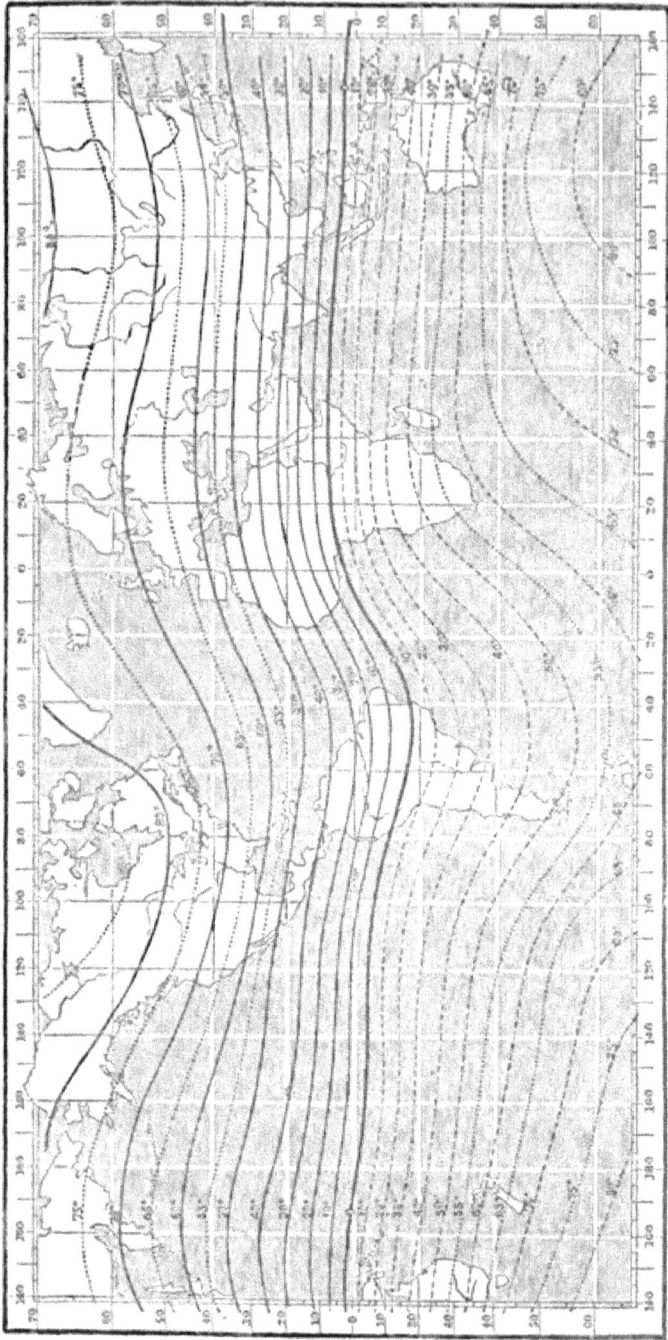

LINES OF EQUAL MAGNETIC DIP.

Inclination is also subject to **secular** change, i.e. its value gradually alters during an extremely long period.

The following table shows the alterations in dip at London :—

Year	Inclination	Year	Inclination
1576	71° 50′	1828	69° 47′
1676	73° 30	1854	68° 31′
1723	74° 42′	1874	67° 43′
1800	70° 35′	1887	67° 26′

Dip may be very well illustrated by

Exp. 51. Suspend a magnetised sewing-needle horizontally by a silk thread (fig. 55). Bring it near a bar magnet, and observe—

(1) at the N-seeking pole of the magnet the needle becomes vertical, with its S-seeking pole downwards ;

(2) move it gradually near the neutral line ; its S-seeking pole gradually rises until

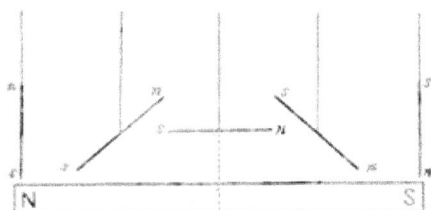

Fig. 55.

(3) it becomes horizontal when its centre is immediately above the neutral line ;

(4) move it gradually towards the S-seeking pole of the magnet ; the N-seeking pole gradually dips until

(5) at the S-seeking pole it becomes vertical, with its N-seeking end downwards.

Exactly similar results would take place if a horizontally suspended needle were carried from the north magnetic pole of the earth to the magnetic equator, and thence to the south magnetic pole. The action of the needle will be understood by reference to fig. 56.

The dotted lines represent the geographical axis and equator of the earth.

Observe that (1) the N-seeking pole dips in the northern hemisphere.

(2) The needle is vertical at the poles.

(3) It gradually becomes horizontal as it approaches the magnetic equator.

(4) The S-seeking pole dips in the southern hemisphere.

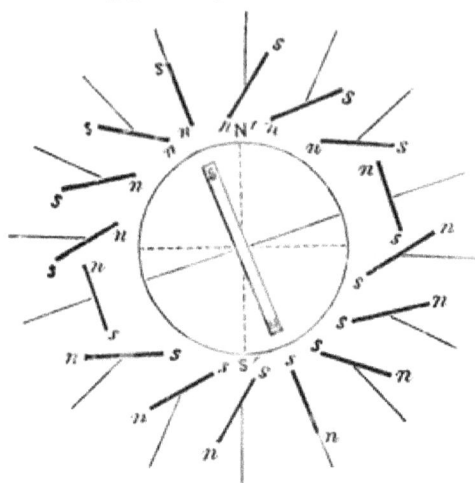

FIG. 56.

The following table shows the Declination and Inclination of various places in the year 1880 :—

—	Declination	Inclination
Boothia Felix . .	None	90° 0′ N. nearly
London	18° 40′ W.	67° 40′ N.
Dublin	22° 30′ W.	69° 42′ N.
Borneo (Labuan) . .	2° 50′ E.	2° 30′ S.
Quito	7° 30′ E.	14° 30′ N.
Cape Town . . .	30° 7′ W.	56° 28′ S.

Magnetic Intensity is the amount of the earth's magnetic force at a place. The *direction* of the magnetic force is the direction taken by the magnetic axis of a magnetised needle when freely suspended at its centre of gravity. It is, however, easier and more convenient to measure what is known as the *horizontal component of the force* by means of a horizontal compass needle. The force which makes such a needle oscillate is only a part of the total magnetic force ;

Thus, let A B (fig. 57) represent the horizontal force,
and the angle B A C   „   „   angle of dip.

On A B construct the rectangle A B C D ; then the diagonal
A C represents the total force.

Taking an easy example : if A B = 1,

and the angle B A C = 60°,

then A C = 2 ;

i.e., when the **dip** is **60°** the total intensity = twice the horizontal intensity.

A method of obtaining the comparative
horizontal force at two places is as follows :—

A horizontally suspended **magnetic** needle
at a place M is moved **from** its position of
rest ; the number of oscillations made in a
given time is counted. If the needle be
made to oscillate at another **place N, it will**
probably **make** a different **number of oscilla-**
tions **in the same time ; e.g., suppose that at**
M it **makes 25 oscillations per minute, and**
at N it **makes 20** oscillations **per** minute.

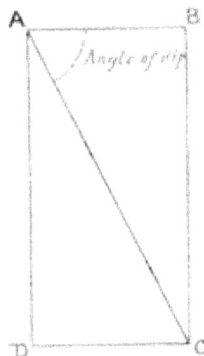

FIG. 57.

Now the force varies as the square of the number of oscillations
at the two places, and we therefore have

$$\frac{\text{Force of earth's magnetism at M}}{\text{Force of earth's magnetism at N}} = \frac{25^2}{20^2} = \frac{625}{400} = \frac{25}{16}.$$

For a further treatment **of this subject** the student must
read a more advanced work on Magnetism.

The Earth's Magnetic Action is merely **Directive.—**
EXP. **52.** Pass a **magnetised** knitting-needle through a cork,
so that **the** ends **project.** Place it in a basin of water. Observe
that the N-seeking pole turns towards the north magnetic pole of
the earth, **but that** there is no movement either to one side of the
vessel or to the other.

We therefore **learn** that the **earth's action** on a **compass-**
needle is not *attractive*, but merely *directive*. This is due to
the fact that **since the size of the** earth is enormously great as
compared with that **of** any artificial magnet, we may consider
that the attractive force exerted by the north magnetic pole of
the earth upon the N-seeking pole of the needle is exactly

equal and opposite to the repulsive force exerted upon the S-seeking pole ; and that the attractive force exerted by the south magnetic pole of the earth upon the S-seeking pole of the needle is exactly equal and opposite to the repulsive force exerted upon the N-seeking pole.

The total effect of these forces upon the poles of the needle is, therefore, equivalent to that which would be produced by two forces equal in magnitude and opposite in direction,[1] one acting towards the north magnetic pole of the earth, and the other acting towards the south magnetic pole. Thus the needle is directed into the magnetic meridian.

The Mariner's Compass.—The magnetic condition of the earth has an extremely important influence on navigation. As early as the twelfth century mention is made in English works of the mariner's compass. As a magnetic needle always points to the magnetic north and south poles, and as declination

Fig. 58.

charts are prepared, the sailor is enabled to guide his ship from port to port by its indications.

The mariner's compass consists of-(i.) a thin, flat compass-

---

[1] Such a system of two equal and opposite forces is called *a couple*.

needle, supported horizontally on (ii.) a pivot of steel, agate, or iridium, which fits into an agate cap ; (iii.) a compass card, consisting of a circular disc fastened *above* the needle, the crown or *fleur-de-lis* being exactly over the north-seeking end of the needle (fig. 58).

The divisions on the card are obtained as follows :—The circle (fig. 59) is divided (1) into four quadrants by means of two diameters at right angles. The extremities of these diameters are marked N, S, E, and W (north, south, east, and west).

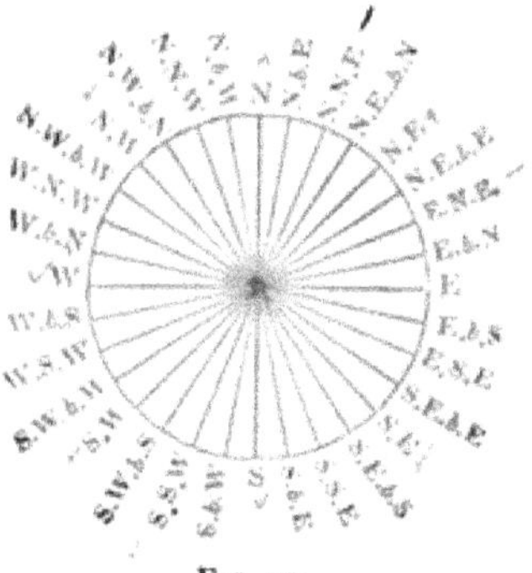

(2) The four right angles thus formed are bisected by lines, the extremities of which form the NE, NW, SE, and SW points. They are named by placing together the two letters at the extremities of the bisected quadrant ; e.g., NE is the point midway between the N and E points.

FIG. 59.

(3) These eight angles are bisected ; the extremities of the lines are named by placing together the letters on the circumference of the bisected angles, remembering, however, to put the name of the cardinal point *first*, and then the name of the other point : e.g., the point midway between the N and NE points is marked NNE (north-north-east) ; that between S and SW is named SSW, and so on.

(4) The sixteen angles are again bisected ; any one of the points thus formed is named by placing (*a*) the name of *one* of the nearest points to it (precedence being given to the point *first obtained* in the above process—i.e., N takes precedence of NNE ; NE takes precedence of NNE or ENE) ; and (*b*) the name of the other nearest cardinal point, the two names being separated by the letter *b* (by) ; e.g.,

The point between N and NNE is marked N *b* E.

,,	,,	NE	,,	ENE	,,	NE *b* E.
,,	,,	E	,,	ESE	,,	E *b* S.
,,	,,	SW	,,	WSW	,,	SW *b* W.
,,	,,	NW	,,	WNW	,,	NW *b* W.

E

It will be seen that there are 32 divisions, which are called *rhumbs*, or *points of the compass.* The angle between two points is $\dfrac{360°}{32} = 11° \; 15'$.

(iv.) The needle and card are enclosed in a cylindrical case, BB', provided with a glass cover (fig. 60).

(v.) The case is supported on gimbals—i.e., two concentric rings, one of which, fastened to the case, moves about two pivots, $x \; d$ (fig. 60). These two pivots are fastened into the

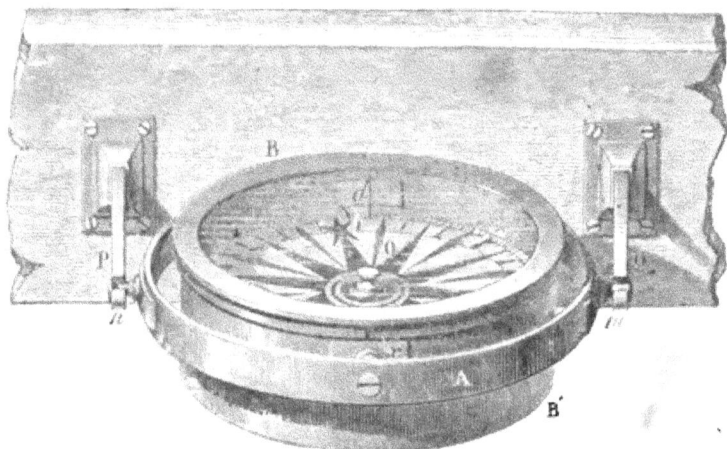

FIG. 60.

other ring A B, which rests by means of the rods m, n on the supports P, Q. By means of the gimbals the compass always remains horizontal, however much the ship may pitch or roll.

If the arrangement is such that the needle moves above the card we have what is commonly called a *land compass*—a useful little instrument for travellers and surveyors.

Astatic Needle or Astatic Pair (fig. 61) is a combination of two magnetic needles arranged so that the earth's magnetism has no directive influence upon them. The two needles are of equal strength and size, one fixed exactly above and parallel to the other, and having their opposite poles in the same direction —i.e., the north-seeking pole a of one magnet is immediately above the south-seeking pole b' of the other. If the strength of two such magnetic needles is the same, the action of the earth on the poles *a* and *b'* and also on the poles *b* and *a'* will be equal and opposite. Such an astatic arrangement will remain in any position in which it is placed.

It is, however, almost impossible for the student to make an **astatic** pair. This will be seen from the following considerations : —

It is difficult (1) to magnetise two needles to exactly the same strength ; (2) to **fix** them parallel ; (3) to **fix** them so that their axes lie in **the same** vertical plane, i.e. so as not to cross one another.

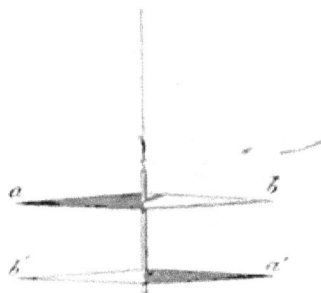

FIG. 61.

A. If the needles **are of unequal** strength, they will, owing to the **action** of the earth's magnetism, tend **to move** into the magnetic meridian.

B. If the **magnets are of** equal size and strength, but their axes are not quite in the same vertical plane, they will set **themselves at right angles** to the magnetic meridian.

Another method of making a needle astatic **is by means of** an ordinary bar magnet.

Exp. 53. Place a **bar magnet** in the magnetic meridian, so that its N-seeking **pole is turned towards** the north magnetic pole of the earth.

(*a*) Bring **a suspended magnetic needle** near it ; the strength of the magnet may overpower the directive action of the earth, and the S-seeking pole of the needle points to the north pole of the earth.

(*b*) Remove the needle farther off ; **the** action of the magnet on the needle is then diminished.

(*c*) By repeated trials place the needle at such a distance from the magnet that its influence exactly counterbalances that of the earth.

EXERCISE IX.

1. People sometimes say, 'The earth is a magnet.' What do they mean ?

2. What is the meaning of the terms 'geographical meridian' and 'magnetic meridian'? What name is given to the angle between the two meridians?

3. I set a rod vertically. When the sun is shining I measure its shadow at any time in the morning, and mark its position ; I also mark the position of the shadow when it is of the same length in the afternoon. I then bisect this angle ; knowing that the line of bisection points to the geographical north pole, how can I find the declination of a place by means of a compass-needle and a quadrant marked in degrees?

4. A sailor observes the position of the N. point of his compass, and then ascertains that the declination or variation of the place he is in is 22° 20′ W. ; what angle must his course make with the magnetic needle so as to steer due west?   *Ans.* 67° 40′.

5. What do you think causes a magnetic needle to set north and south ? Does its direction ever vary? If so, state what you know regarding the variation.

6. Supposing the top of the page on which you are writing to represent the north, and the bottom to represent the south, you are required to draw the true geographical meridian, then to draw a magnetic needle in the position which it occupies with reference to that meridian, and to mark on your figure the angle of declination and the number of degrees it embraces in our latitude. Why do I use the words ' our latitude ' ? Give also a sketch of the dipping-needle, and name the angle that it makes with the horizon.

7. How would you illustrate the fact that the N-seeking end of a magnetised needle dips downwards, having given a magnet, a knitting-needle, thread, and a file.

8. What are meant by the declination and inclination of the magnetic needle ? What are meant by the terms magnetic dip, magnetic poles, and magnetic equator ?

9. If you were required to make a model to illustrate the magnetic properties of the earth by putting a bar magnet inside a ball of clay, show by a sketch how you would place the magnet, and explain how the magnetic properties of the model would answer to those of the earth.

10. A long strip of hard steel is magnetised, and when a small magnetic needle is passed along the strip, its north point is attracted by one end of the strip, its south point by the other, the centre of the strip appearing to attract neither point of the needle. When the strip is broken across at the centre, what is the action of its two halves upon the magnetic needle ?

11. What is meant by saying that the magnetic dip at London is 67° 30′ ? State in general terms at what places on the earth's surface the magnetic dip is least.

12. It is sometimes said that the earth has no tendency to impart to a magnetic needle a motion of translation, but that it has under certain circumstances a tendency to impress upon it a motion of rotation : what is the meaning of these statements ?

13. How could the position of the magnetic north pole of the earth be

discovered by means of a magnetised needle suspended at its centre by a piece of silk?

14. You carry with you a dipping-needle from the earth's north magnetic pole across the equator to the south magnetic pole; how will the dipping-needle behave during the excursion?

15. State what you know regarding the use of the magnetic needle in the steering of ships.

16. What are meant by isoclinic and isogonic lines?

17. Why do we apply the **term** north-seeking to the pole of a magnet which points to the north magnetic pole of the earth?

18. Two equal bars of steel, after having been equally magetised, are kept for **some** years in **a** vertical position, one (*a*) with its south-seeking pole **upwards, the** other (*b*) with its north-seeking pole upwards. The bars **are so** far **apart** that they **do** not act on one another; which of the two bars would you expect to find had kept its magnetism best, and why?

19. Being provided with **a** dipping-needle only, **how could** the magnetic meridian be found?

20. What are meant by diurnal, annual, **and secular** variations of a magnetic needle?

21. What is meant by **a** magnetic storm? **How** do we know when a magnetic storm occurs?

22. **How is** the strength **of** a freely suspended magnetic needle kept up?

23. **How would** you construct an astatic needle out **of a** uniformly magnetised strip of watch spring, which you are allowed to **bend** or break as you please?

24. A strong bar magnet **is placed with** its axis lying in the magnetic meridian, and with its N-seeking **pole towards the** north. State in what direction a compass-needle points, (1) when placed immediately over the centre of the bar magnet; (2) when gradually raised vertically upwards.

(N.B. The compass-needle **can** only turn about its pivot **in a** horizontal **plane.**)

25. **An astatic** combination **of** two magnets is injured, so that the magnets **are at** right angles instead **of** parallel to each other. If it be suspended **as** usual, what position will it assume with regard **to** the magnetic meridian? Illustrate your answer with a diagram showing the forces **which act upon the magnets.**

FRICTIONAL ELECTRICITY.

CHAPTER VI.

ELECTRICAL ATTRACTION AND REPULSION.

Electrical Attraction.—Exp. 54. Warm and dry [1] a glass rod or a stick of sealing-wax. Rub the glass rod with a piece of warm silk, or the sealing-wax with a piece of warm flannel. Bring the rubbed rod near some bits of paper, straw, bran, or any other light bodies. Notice that these bodies are attracted by the rod (fig. 62).

Thus by rubbing, an additional and curious property has been imparted to the rod. This power of attraction was originally recorded more than 2,500 years ago by Thales, a Greek philosopher (600 B.C.), who mentioned that rubbed amber would attract light bodies.

FIG. 62.

[1] **Moisture** on the surface of apparatus used in frictional electricity is fatal to successful experimental work. In order that experiments may be satisfactorily performed, both the rubber and the body rubbed must be quite dry. It is advisable to place all rods, rubbers, and apparatus in front of an ordinary coal fire, or a gas reflecting stove for some time before use. Unless this precaution is adopted much valuable time will be wasted. It ought to be remarked that experiments sometimes fail because rods are handled too much. This arises from the presence of moisture on the skin.

The Greek word for *amber* is ἤλεκτρον (electron), from which we have derived our word *electricity*.

We will now perform a few experiments to further illustrate the phenomenon of electrical attraction.

Exp. 55. Take a small pith ball, attach to it a piece of *cotton* thread, and suspend it on a suitable support, as shown in fig. 65. (Elder pith is best for this purpose. After cutting the ball into shape with a sharp knife, it is advisable to press it slightly between the fingers so as to remove any projecting points.) Bring near a warm and dry glass rod rubbed with warm silk, or a dry rod of sealing-wax rubbed with warm flannel. Notice that the pith ball is attracted to the rod.

When bodies—e.g., glass rubbed with silk, or sealing-wax rubbed with flannel—exhibit this power of attraction, they are said to be *electrified*, *excited*, or *charged*.

Exp. 56. Place an empty egg-shell on a table ; bring an electrified rod near. Notice that the shell rolls after the rod as the latter is gradually withdrawn.

Exp. 57. Balance a lath (say 6 ft. long) upon the bottom of an

Fig. 63.

inverted glass flask (fig. 63). Hold an electrified glass rod near one end. The lath is attracted.

Exp. 58. Rub the smooth face of a piece of resin on hot dry flannel. Use the resin instead of the glass rod in the last experiment.

Exp. 59. Attract the lath by a rod of shellac or of sulphur rubbed with flannel.

Exp. 60. Brush a dry and *hot* sheet of brown paper with a clothes-brush :—

(*a*). Hold it near the wall of a room : it will immediately cling to the wall, and will continue to do so for some time.

(*b*). Remove the paper, brush again, fold, and present it to a balanced lath. Notice attraction.

(*c*). Hold it over light bodies as in Exp. 54 ; they are immediately attracted.

(*d*). Hold it over a boy's head : many of the hairs will stand upright.

why?

Exp. 61. Heat a drawing-board and place warm foreign note-paper upon it ; rub the paper with bottle india-rubber. Remove the paper and place it on the other side of the board, or on a wall. Notice how it clings.

The paper may also be used to attract the balanced lath, or any light body.

Exp. 62. Place pieces of vulcanised india-rubber over two fingers : strongly electrify silk ribbon by drawing it between them. Notice how it attracts a lath, or clings to a wall.

Exp. 63. Bend a piece of wire, and fasten it to a *silk* thread as shown in fig. 64. Electrify a piece of sealing-wax or a glass rod. Place it across the stirrup. If now the hand be held near, the rod will be attracted.

It should be remarked that the attraction in these experiments is mutual ; in all cases the electrified body attracts, and is attracted by, the unelectrified body.

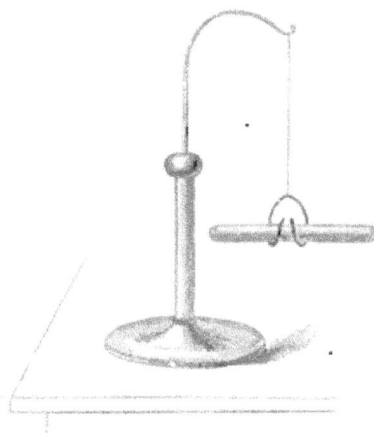

FIG. 64.

EXERCISE X.

1. What is the derivation of the word *electricity*?

2. Give a number of experiments you can do with a glass rod and a piece of silk.

3. Mention any substances which can be electrified by rubbing them with flannel.

4. How would you show that an electrified body is attracted by an unelectrified one ?

Electrical Repulsion.—Exp. 64. Repeat Exp. 54, and notice that after the light bodies have been in contact with the rod they are for the most part immediately repelled.

To observe this repulsive action more readily take a small

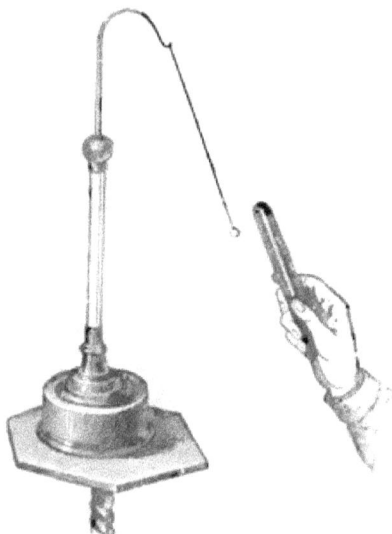

FIG. 65.   FIG. 66.

pith ball and suspend it by a fine *silk* thread from a suitable support, which may be made of glass, metal, or wood (figs. 65 and 66). This apparatus is called the *Electric* or *Pith-ball Pendulum*.

Exp. 65. Place the pendulum before a fire to dry the silk. Bring an excited rod near. Attraction first takes place (as in Experiment 55), but after contact with the rod the ball is violently repelled, and will not again approach the rod (unless the ball has come in contact with the wooden or metallic stand).

A similar experiment may be performed by using a feather instead of a pith ball.

From this experiment we learn two lessons :—

(1) That a body becomes electrified, not merely by friction, but by touching it with an electrified body.

(2) That **if one body be** electrified by **contact** with another body, they **repel one another.**

Exp. 66. Electrify foreign **note-paper** as in Exp. 61 ; cut the paper into strips, leaving about **an inch** margin. **Roll the uncut** part up and hold it in one hand (fig. 67). Notice that the strips repel one another, because they are charged with **electricity** from **the same source.**

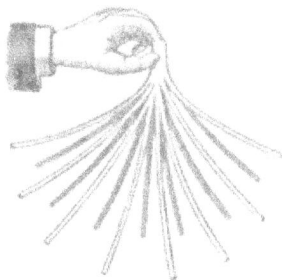

Fig. 67.

Exercise **XI.**

1. Two pith balls are suspended by silk threads. They are both charged with electricity by means of a vulcanite rod rubbed with flannel. They are then brought side by side. What happens ?

2. A stick of sealing-wax is rubbed with dry flannel and held over a pith ball lying on a table. The ball rises to the sealing-wax and then falls again. Why does it rise, and why does it fall ?

Other Properties **of Electrified Bodies.—Exp. 67.** Strongly electrify a glass tube and hold it near the face. A peculiar sensation, likened to that of coming in contact with cobwebs, is perceived.

Exp. 68. Vigorously excite a warm vulcanite rod with a warm piece of flannel ; a '**crackling**' is heard.

Thus, **not** only do **bodies give rise to the** phenomena **of (1) attraction and** (2) repulsion ; **but there are** produced **(3) a peculiar sensation** if **the rod be held near the** face ; (4) **a crackling ; (5) under certain** conditions, luminous effects ; **and (6)** a faint **peculiar odour, due to a** gas called ozone. The last **two** effects **are especially noticeable** when an electrical machine is worked.

Two kinds **of Electrification.—**We **have learnt by experiment** that **if an** electrified **rod be brought near a pith ball,** the ball will **(1)** be attracted, **(2) become** charged by contact, and **(3) repelled. One of the lessons learnt was that** bodies electrified from the same **source** repel each **other.**

Exp. 69. Now **charge the** pith ball **with electricity obtained** from a glass **rod** rubbed with silk. Remove **the glass rod and bring** near a **piece of** sealing-wax rubbed **with flannel. The pith ball** immediately rushes **to** the sealing-wax.

Exp. 70. Suspend (as in fig. 64) a glass rod rubbed with silk.

1. Vigorously electrify another glass rod, and hold it near the suspended rod. Repulsion takes place.

2. Bring up a sealing-wax rod rubbed with flannel. Notice attraction.

If a sealing-wax rod rubbed with flannel be suspended, the electrified glass rod attracts, and an electrified sealing-wax rod repels it.

These experiments indicate that there are two kinds of electrification, viz. :

    (*a*) that from glass rubbed with silk ;

    (*b*) that from sealing-wax rubbed with flannel.

They also show that if a body be charged with one kind of electrification, and another body be charged with the other kind of electrification, they mutually attract ; but if they are both charged with the same kind they repel one another.

Electrification derived from glass rubbed with silk was formerly called *vitreous* (from *vitrum*, the Latin word for glass); that derived from sealing-wax or resin rubbed with flannel was called *resinous*. These names, however, have long been superseded by others, as it was found that vitreous electricity could be obtained from substances which produced resinous electricity, and *vice versâ*, by merely altering the material of the rubber.

Vitreous electricity is now commonly called Positive or plus (+) electricity.

Resinous electricity is now commonly called Negative or minus (−) electricity.

These terms are arbitrary ; there is no physical reason for calling one kind positive rather than the other.

We are now in a position to state the previous results as follows :—

A body charged with positive electricity repels a body charged with positive electricity ;

A body charged with negative electricity repels a body charged with negative electricity ;

A body charged with negative electricity attracts a body charged with positive electricity.

Or shortly the first **law** stands :—

(*a*) Bodies **whose** electrifications are of the same kind repel.

(*b*) Bodies whose electrifications are of unlike kinds attract.

Theories of Electricity.—Thus **far** we have dealt with certain phenomena, but the questions often arise, **What is** electricity? **How** can we account for these phenomena? **To** answer these questions several **theories** have been **advanced,** none of which are quite satisfactory. The suppositions, however, are **aids** by which we may account for the great facts of electrical **science, some** of which have already been brought before your notice.

I. The two fluid [1] **theory** was originated **by Symmer. He** supposed that all unelectrified bodies were possessed of two electric **fluids of opposite kinds called positive** and negative, which were, **(1)** equal in quantity, (2) inexhaustible, (3) without weight, (4) self-repulsive, **i.e.** each **fluid repels its** own kind, (5) mutually attractive, **i.e.** each **fluid attracts its** opposite kind.

According **to** this theory, a **neutral body is one in** which the **two** fluids neutralise each **other, i.e. are present in equal** quantities. By friction (or **other causes) the two fluids are partially** separated, so that **the positive fluid goes to one body** and the negative fluid to the other.

A glass rod, **for example,** is neutral **in its ordinary state,** because it possesses equal quantities of the **two fluids ; a piece** of silk is neutral for a similar reason.

If **they are rubbed together, the** glass **rod** receives an excess of positive, the silk **receives an** excess **of negative.**

II. **The one fluid theory** of Franklin supposes that all unelectrified bodies possess **a** single fluid, which is, (1) **without** weight, **(2)** self-repulsive, (3) diffused in definite quantities **through all** bodies.

According **to this theory** bodies are electrified **by friction**

---

[1] The term ' fluid ' used in these theories is misleading. The so-called electrical fluid and a material fluid differ in very many properties. The chief point of resemblance between the two lies in the fact that both are capable of flowing from one point to another according **to similar** laws.

when the fluid passes from one body to the other, so that it is in excess on one body and is deficient on the other. The body which has more than its share is positively electrified, that which has less is negatively electrified.

III. The **molecular theory** supposes that electrification is due to peculiar movements of the molecules of the body, or of the ether surrounding the molecules. ·

The last theory is probably the correct one ; at any rate, whatever the nature of electricity may be, it has been amply proved that the mysterious fluid does not exist.

Simultaneous and equal development of the two kinds.— **Exp. 71**; devised by Faraday. Place a flannel cap, to which a silk thread is attached, over one end of a stout rod of shellac or vulcanite, both having been previously warmed (fig. 68). Rub the cap round the rod a few times.

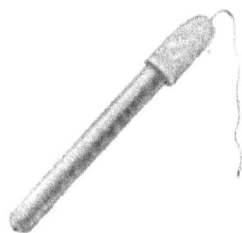

1. Remove the cap by the silk thread, and hold it near a positively charged pith-ball pendulum. **Notice** repulsion. The flannel is, therefore, **charged** positively.

Fig. 68.

2. Present the rod to the pith ball ; violent attraction ensues. It is, however, preferable to charge the pith ball negatively. **Notice** repulsion ; therefore the. rod is negatively electrified.

3. Replace the cap and rub again. Without removing the cap hold both to an uncharged pith-ball pendulum. There is neither attraction nor repulsion.

This experiment is best performed with a gold-leaf electroscope (described on p. 66), instead of a pith-ball pendulum.

Thus we conclusively **prove** that (1) positive **and** negative electrifications are generated together ; in fact, one kind of electrification is never produced without the other ; and (2) the positive electrification is exactly equal in amount to the negative.

Bodies not absolutely positive or **negative.—Exp. 72.** Electrify a pith-ball pendulum by means of sealing-wax rubbed with flannel (*i.e.* negatively). Excite a very hot glass rod with cat-skin or fur, and bring it near the pith-ball. **Notice** repulsion. The glass has therefore had negative electrification produced upon it.

The kind of electrification developed by friction thus depends not only upon the body rubbed, but **also** upon the rubber.

List of Bodies in which, any two being rubbed together, **the** one standing first becomes positive, the other negative.

Fur	Caoutchouc
Flannel	Sealing-wax
Ivory	Resin
Glass	**Amber**
Cotton	Sulphur
Paper	Gutta-percha
Silk	Collodion
The hand	Gun-cotton
Wood	Leather coated
Metals	with amalgam

Thus if **we choose** silk and resin, silk becomes positive **and** resin negative. If we choose **leather** (covered with amalgam) and resin, the leather becomes negative and the **resin** positive.

Exp. 73. Test as many of these bodies as possible by means of a pith-ball pendulum, **electrified from a known source.**

It must be stated **that the results** of experiments **on bodies** standing close **together in the list** are somewhat uncertain, as a slight difference **in** chemical composition, or in the nature of the surfaces, **may** alter their behaviour.

Exp. 74. Rub a piece of hot smooth glass on a piece of hot rough glass. Show that the rough glass becomes negative, the smooth positive.

Exp. 75. Take a piece of silk ribbon, cut it into two parts, **rub** one *across* the other. **Test that the one** rubbed crosswise becomes negative, **the other positive.**

As a general rule, if two similar bodies are rubbed together the one whose particles **are** most easily displaced becomes negative.

Sources of Electricity.—Electricity is generated in various ways, which may be **tabulated thus :—**

Source		Name
Mechanical : e.g. friction pressure concussion cleavage mere contact	}	Frictional or statical electricity
Chemical :		Voltaic, galvanic, or dynamical electricity
Physical :	Heat	Thermo-electricity
	Magnetism	Magneto-electricity

The electricity produced by these sources is similar in their *nature*, but different in *intensity* and *quantity*.

EXERCISE XII.

1. How did the terms *vitreous* and *resinous* arise? Describe an experiment to prove that they are unsuitable names for the two kinds of electricity.

2. Give an experiment to show that positive and negative electrifications are produced together.

3. What is the relation between the amount of electricity on the body rubbed and on the rubber?

4. Mention the sources from which electricity can be produced.

5. When a piece of sealing wax and a piece of dry flannel are rubbed together, one becomes positively electrified and the other negatively electrified. When a piece of dry paper and a piece of india-rubber are rubbed together, one becomes positively electrified and the other negatively electrified. How could you find out which of the four things—sealing-wax, flannel, paper, india-rubber—are in the same electrified state?

6. You are required to prove by experiment the electrical law that bodies similarly electrified repel, and that bodies dissimilarly electrified attract each other; how would you do it?

7. A sheet of hot paper is placed on a hot board, and india-rubber is passed briskly over it. Two strips are cut from the paper and held up close and parallel to each other. How will they act upon each other? A glass rod, rubbed with silk, repels both the strips; what is the inference?

CHAPTER VII.

ELECTROSCOPES

IT is necessary now to understand the construction of certain instruments which are used to detect the presence, and determine the kind, of electrification of a body. **These** instruments are called **Electroscopes.**

I. **The first** electroscope was **used** as early as 1600 A.D. by Gilbert, and **consisted** of a straw balanced on a **fine** point.

Fig. 69 represents a straw delicately supported on the point of a sewing-needle. The point **fits into a cap of straw** stuck

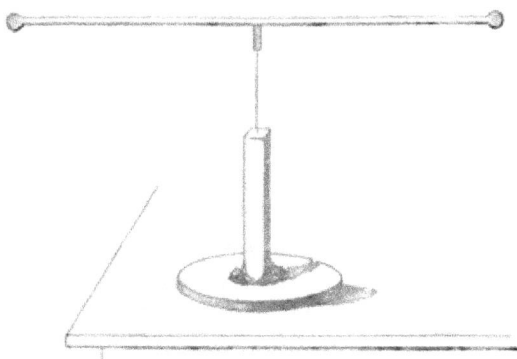

FIG. 69.

on with sealing-wax. The other end of the needle **is inserted** into a stick of sealing-wax attached to a circular plate.

II. **The Pith-ball Pendulum** has been previously explained (p. 58).

III. **The Pith-ball Electroscope** (fig. 70) consists of two small balls of elder pith, suspended by *cotton* threads from one

F

end of a brass wire, the other end being terminated by a knob. A hole, somewhat larger in diameter than the wire, is bored through a cork, which fits into the mouth of a well-dried glass vessel. The wire is inserted through the hole in the cork, which is then filled up with hot shellac.

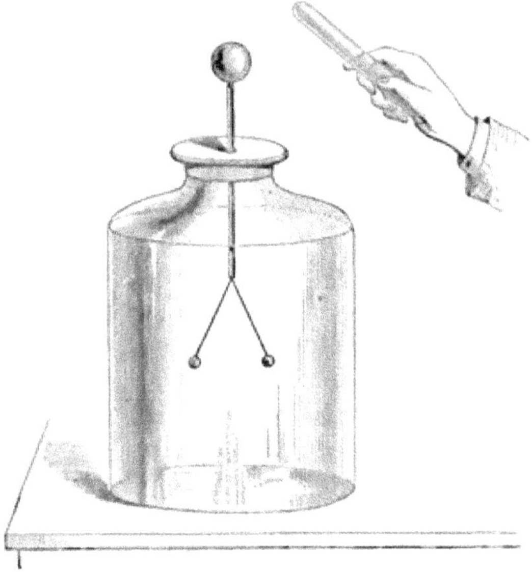

FIG. 70.

IV. **The Gold-leaf Electroscope** (fig. 71) is similar in principle to the pith-ball electroscope just described. A metal ball, or preferably a disc with well-rounded edges, is fixed to a brass rod which passes through a stout varnished glass tube fixed into the dry wooden cover of a wide-mouthed glass vessel.

The other end of the rod has a horizontal flat cross-piece (say, three-quarters of an inch long) fastened to it, on each side of which a piece of gold leaf (two inches long) is gummed. The leaves are thus parallel and hang close together. On opposite sides of the interior of the vessel are sometimes placed two strips of tinfoil, of sufficient height to be touched by the gold leaves on their extreme divergence. They are so placed as to be in connection with the earth. If these strips are not present the leaves are liable to be broken by contact with the sides of the vessel.

To keep the interior dry it is advisable to place therein a vessel containing small pieces of calcium chloride, or a few fragments of pumice-stone soaked in strong sulphuric acid. A gold-leaf electroscope, when perfectly dry, will indicate exceedingly small charges of electricity.

FIG. 71.

To make a Gold-leaf Electroscope.—At the Normal School of Science, South Kensington, the students are instructed to make an electroscope as follows :—

'Clean and dry a glass flask; fit it with a cork. Bore a half-inch hole through the cork. Solder a penny piece or disc of metal to one end of a straight brass wire seven inches long. Drill a hole in the edge of the disc. Surround the wire about two inches below the disc with shellac, so as to fit the hole in the cork. Solder a piece of thin sheet brass about half an inch long to the lower end of the wire. Cut two strips of gold leaf half an inch wide. Slightly moisten the brass strip on each side with weak gum and take up the leaves. Shade from air currents and place in the flask.'

Uses of a Gold-leaf Electroscope.— The electroscope is used—

(1) to indicate the *existence* of electricity,

(2) to determine the *kind* of electricity,

(3) to indicate, very roughly, the *amount* of electricity on a body.

We will now show these uses by a series of experiments.

(1) To indicate the existence of electricity on a body.

Exp. 76. Rub a rod of sealing-wax or vulcanite with flannel, and bring it gradually near the electroscope. Notice that the leaves diverge.

Exp. 77. Strike the disc with fur or with a fox's brush.

Exp. 78. Grind sulphur to powder in a mortar, and drop a little on the disc.

Exp. 79. Grind warm coffee into a hot glass tumbler and sprinkle it on the disc.

(2) To determine the kind of electricity generated on a body.

Rule.—If, when an electroscope is charged, the approach of an electrified body causes the leaves (*a*) to diverge *more*, the approaching body and the leaves are similarly electrified ; (*b*) to diverge *less*, the approaching body and the leaves are oppositely electrified ; or else the approaching body is uncharged.

This rule will be understood after reading Chapter IX.

Perform the experiments shown in the following table.

For the third column (+ charge) use a glass rod rubbed with silk.

For the fourth column (– charge) use a vulcanite rod rubbed with flannel.

For the fifth column (uncharged body) use the hand.

Care must be taken that the charged body is brought up *cautiously*.

Compare the results in the following table with the list on p. 63.

It is important to notice that the *collapse* of the leaves does not necessarily indicate that the body to be tested is charged ; they may collapse if the body presented to the electroscope is an uncharged conductor.

Number of experiment	Electrified body to be tested	Bring up a positively charged body	Bring up a negatively charged body	Bring up an unelectrified conductor	Inference
80	Beat the disc with a fox's brush	the divergence is less	the divergence is greater	the divergence is less	The disc is negatively electrified
81	Sulphur powdered in a mortar and dropped on the disc	less	greater	less	Sulphur is negatively electrified
82	Flannel cap rubbed on vulcanite rod (as in Exp. 71) and the cap placed on the disc	greater	less	less	Flannel cap is positively electrified
83	Sealing-wax rubbed with flannel and placed on the disc	less	greater	less	Sealing-wax is negatively electrified
84	Coffee ground and dropped on the disc	greater	less	less	Coffee is positively electrified

Repulsion is the only sure test of Electrification—From previous experiments we learnt that there is mutual attraction between electrified and unelectrified bodies; we must, therefore, not rely upon attraction to indicate a definite kind of electrification. For suppose we have a pith-ball pendulum charged positively, and we have to ascertain whether a rod is charged or not. Attraction takes place with a negatively charged rod or an uncharged rod, but repulsion can take place only if the rod is charged positively.

(3) A gold-leaf electroscope will also *very roughly* indicate the amount of electrification of a body.

Exp. 85. Hold a slightly electrified rod near the disc of an uncharged gold-leaf electroscope. There is only a slight divergence of the leaves.

Exp. 86. Present a highly charged rod. Notice greater divergence, even when the rod is held much farther from the electroscope.

Instruments by means of which the quantity of electricity imparted to bodies can be measured are called **Electrometers.**

V. **Henley's Quadrant Electrometer.**—This little instrument is really an *electroscope,* for although it roughly *measures* the quantity, its real use is to *detect* the presence of large charges of electricity.

It consists of an upright brass or wooden rod *d*, to which is attached a graduated scale. In the centre of this is a small index, made of light wood or straw carrying a pith ball. It is generally used to ascertain whether an electrical machine is charged, and in fig. 72 it is shown fitted into the prime conductor of a machine. The pith ball at first hangs vertically downwards, and as the prime conductor is charged the pith ball is repelled. The reason of this is that the upright rod and the ball are similarly electrified.

FIG. 72.

EXERCISE XIII.

1. How would you make a gold-leaf electroscope? Give a diagram.

2. Describe an experiment to prove that repulsion is the only sure test of the presence of free electricity.

3. A collodion balloon simply stroked with the hand becomes negatively electrified. Supposing you were asked to prove the truth of this statement, how would you proceed?

4. How would you prove that the electrification of sealing-wax rubbed with flannel is similar to that of amber rubbed with flannel?

CHAPTER VIII.

CONDUCTION.

Electrics and Non-Electrics.—Exp. 87. Rub a brass rod, held in the hand, with warm silk. Touch the disc of an electroscope with the rod. The leaves are unaffected.

Exp. 88. Mount the brass rod on a glass or ebonite handle (fig. 73), or hold it with a sheet of gutta-percha. Thoroughly dry the handle and rub as before. Bring it near the disc of the electroscope : notice the divergence of the leaves.

Fig. 73.

Exp. 89. Repeat the last experiment, but before bringing the rod near the disc touch it with the finger. The leaves no longer show signs of electrification.

For many years it was imagined that only a certain number of bodies could be electrified ; these were called **electrics**, to distinguish them from bodies which, it was imagined, could not be electrified, as they exhibited no signs of acquiring the property of attracting light bodies even after violent friction. These bodies were termed **non-electrics**. This distinction was founded upon a scientific error, as all bodies are capable of being electrically excited, though in different degrees, if proper precautions are adopted.

From the last three experiments we learn that a rubbed brass rod, when held in the hand, or when touched with the finger, shows no evidence of electrification, but that when mounted on a glass handle the electrification becomes at once apparent. The reason of this escaped the early observers. The fact is, the electricity is discharged through the hand and body of the experimenter to the earth in the former case, but it is not so discharged in the latter.

Conductors and Insulators.—Bodies which thus readily allow electricity to spread over them, and to pass away to other bodies, i.e. bodies which offer little *resistance* to the flow of electricity, are called *conductors*. Those which offer great resistance to its passage are called *non-conductors, insulators,* or *dielectrics.*

In the three experiments just performed the brass is a conductor and the glass an insulator.

The discovery that some bodies conducted electricity, but that other bodies did not conduct, was made by Stephen Gray in 1729. He noticed that a cork inserted into a rubbed glass rod attracted light bodies. He then stuck pieces of wood into the cork, and still light bodies were attracted. He also found that pack-thread, 765 feet long, if suspended on *silk* loops, would allow electricity to pass through it, but if *wire* loops were used the electricity would no longer pass. He arrived at the conclusion that silk prevented, and wire allowed, the flow of electricity to the earth—in other words, his conclusions were that silk was a non-conductor and wire a conductor.

Exp. 90. Insert a cork into a glass tube, rub the tube close to the cork, and notice that light bodies are attracted when the cork is presented to them.

Exp. 91. Insert rods of wood (varying in length from six inches to three feet) into the cork ; rub the glass as before : the free end of the wood will attract a balanced lath.

Exp. 92. Tie a thread of *cotton* round one of the rods in the last experiment, the free end of which has a metal ball or key attached to it. Electrify the tube. Small bits of paper will be attracted by the ball (fig. 74).

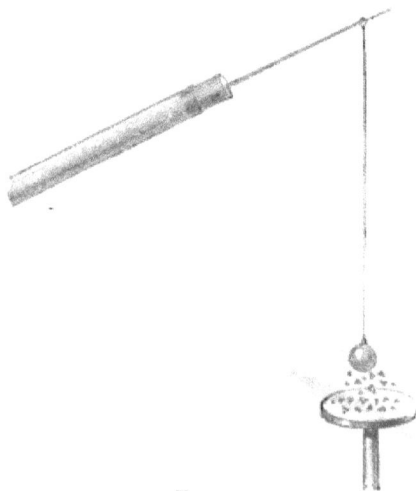

FIG. 74.

Exp. 93. Try the last experiment with a silk thread. The bits of paper are not attracted.

Wood and cotton are, therefore, conductors : silk is a non-conductor.

The conducting power of various bodies may be excellently shown by the following experiment :—

Exp. 94. Place (*a*) one end of a long fine copper **wire** through the hole in the disc of the electroscope. Make a loop at the other end and place it over a vulcanite rod. Rub the rod with flannel and allow the loop to slide down **the rod.** Notice divergence of **leaves.**

   Replace wire by (*b*) cotton thread (divergence).
      ,,   ,,   (*c*) dry silk thread (no action).
      ,,   ,,   (*d*) wet silk **thread (divergence).**

Exercise XIV.

1. What inference do you draw from the results **in Exp. 94?**

2. A pith ball **is suspended from** a metal stand **by a fine** thread. **If** you have a strongly electrified glass rod, **how** can you find **out** whether the thread is a conductor or a non-conductor of electricity?

3. A stick **of** sealing-wax, held in **the hand and rubbed with dry** flannel, is found **to be electrified.** A brass rod after **being** treated in **the** same way shows no **electrification.** How do you **account** for this difference?

4. How **would you prove** that a brass **tube** is negatively electrified when struck **with flannel?**

5. Define the terms *electrics* and *non-electrics.* Why was the distinction made, **and why has it been discarded?**

There **is no hard and fast division** between conductors and non-conductors, as (**1**) **many** substances are partial conductors, i.e. **they** neither allow **the** electricity to flow readily nor offer any **great resistance to it, while (2)** all bodies offer some resistance to the passage **of electricity.** **When** this resistance is almost inappreciable **the body is a** *good conductor*; **when the** resistance is very **great, practically no electricity passes from one** point to another; **the body** is then a *non-conductor*. We therefore learn **that the term** *conductor* is merely relative.

List of Conductors *in the Order of their Conducting Power, those standing first on the list being the best conductors, and those at the bottom being the best non-conductors.*

Silver
Copper
Other metals
Charcoal
Sea water   } Good conductors
Pure water
The body
Cotton

Dry wood
Marble
Paper   } Partial conductors
Straw

Oils
Porcelain
Wool
Silk
Resin
Gutta-percha   } Non-conductors or insulators
Sulphur
Shellac
Ebonite
Paraffin
Dry air

The fact that water is a good conductor enables us to understand why electrical apparatus should be quite dry.

Exp. 95. Charge a gold-leaf electroscope by contact. Touch the disc with various bodies in the above list, recharging when necessary. Notice that with

(*a*) a good conductor the leaves collapse instantaneously;
(*b*) a non-conductor the leaves are not perceptibly altered.

EXERCISE XV.

1. What influence has moisture in electrical phenomena? Give instances.

2. How is it that in damp weather an ordinary plate machine will not work well?

3. Give the reasons of the results obtained in Exp. 95.

4. Arrange the following substances in the order of their conducting powers for electricity, putting the name of the best conductor first :—air, copper, glass, iron, sea water, shellac, water (pure), wood.

5. How would you ascertain whether a stone is a conductor or not?

6. A distant insulated sphere is connected in succession with a gold-leaf electroscope by means of

        (1) a metallic chain,
        (2) dry silk,
        (3) india-rubber,
        (4) a wire,
        (5) moistened cotton,
        (6) moistened silk.

An electrified glass rod is held near the sphere. What happens to the leaves in each case? Give reasons.

Insulation.—To insulate a body it must be supported on, or suspended by, a non-conductor.

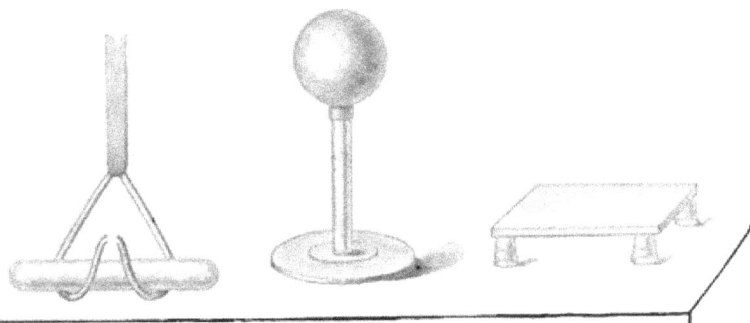

FIG. 75

The most usual forms of insulators are—

(1) Stems of glass coated with shellac varnish,[1] ebonite, or vulcanite.

(2) Sheets of vulcanised india-rubber, ebonite, or varnished glass.

(3) Dry threads of silk.

[1] Pour one ounce of methylated spirit on one ounce of shellac, and let it remain for twenty-four hours. Then dilute it with three ounces of spirit. Strain it through flannel. To varnish apparatus, first warm them, then draw the brush regularly in the same direction ; allow the varnish to dry. Repeat this process if necessary.

Common methods of insulation are shown in fig. 75. The wire stirrup is suspended by a silk ribbon or thread; the spherical conductor is supported on a varnished glass stem. The insulating stool, on the right of the figure, is made by placing a mahogany board on four strong, varnished glass tumblers, or on varnished glass legs.

Proof Plane.—To obviate the inconvenience and difficulty of testing large charges of electricity, a small instrument, called a proof-plane, is exceedingly useful (fig. 76). It consists of a small conductor supported on an insulating handle.

An excellent one is made from a disc of metal, an inch or so in diameter, fastened to a rod of sealing-wax. This can easily be made, especially with a soft metal such as zinc. A card covered with tinfoil fastened to a handle of ebonite, sealing-wax, shellac, or varnished glass answers well. A conductor of any shape may be used, provided that the corners and edges are thoroughly rounded off. (The reason of this will be understood after the 'Action of Points' has been read.)

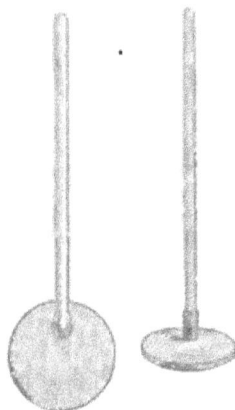

Fig. 76.

Sometimes the conductor consists of a small brass ball suspended by a silk thread. It is then commonly called a **carrier ball**.

When a charged body is touched with a proof-plane the two form, electrically speaking, one body; on separating them a small quantity of electricity is removed on the proof-plane, which can then be tested without danger of fracturing the leaves of an electroscope.

EXERCISE XVI.

1. What is meant by the terms *conduction* and *insulation* as applied to frictional electricity? Describe an experiment which shall illustrate the properties of metal wire, common twine, and a silk string as regards conduction and insulation.

2. Two pith balls hang side by side by two damp cotton threads.

State and explain what happens when an excited glass rod is brought gradually near the two balls from below.

3. Gutta-percha and silk are insulators. How would you ascertain which is the better for use under water?

4. You have several rods of unknown materials. Describe exactly experiments which would enable you to distinguish those which are conductors of electricity from those which are non-conductors.

5. What occurs if I hold an uninsulated brass plate between a gold-leaf electroscope and a charged vulcanite rod?

6. Why is it that the electricity generated on a glass rod does not escape when the rod is held in the hand?

7. An apple held in the hand and struck with a fox's brush shows no signs of electrical action; suspended by a string of silk and struck with the brush it becomes electrified, attracting light bodies, and causing the leaves of an electroscope to diverge. Explain these results.

8. Small particles of paper lying on a table are acted upon by an electrified vulcanite rod. Small and light particles of india-rubber are not so acted on. Explain this.

9. Supposing you were required to test the quality of the electricity with which an insulator is charged, how would you do it?

10. Describe an experiment showing that the human body is a conductor of electricity.

11. A rod of sealing-wax is rubbed with dry flannel. An uncharged pith ball suspended by a silk thread is attracted when the sealing-wax is brought near it, but is unaffected by the flannel. Would you conclude from this experiment that when sealing-wax and flannel are rubbed together the sealing-wax only is electrified? Give reasons for your answer.

CHAPTER IX.

INDUCTION.

THE early experimenters soon discovered that bodies could be charged otherwise than by direct contact.

Exp. 96 (due to Stephen Gray). Support a lath on a warm varnished tumbler (fig. 77). Place small pieces of paper under

FIG. 77.

one end of the lath. Hold a strongly electrified rod over the other end, but not in contact. Notice that the paper is attracted.

Exp. 97. The last experiment can be varied in an interesting manner by placing a gold-leaf electroscope under one end of the

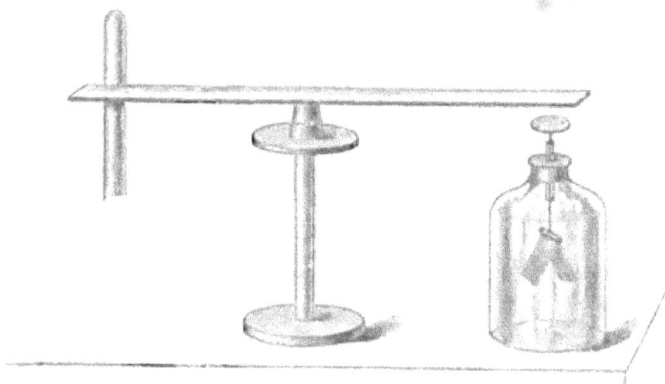

FIG. 78.

lath. Bring a strongly electrified rod near the other end. The gold leaves diverge (fig. 78). Attract the lath so that it moves round. Notice that the gold leaves diverge every time the lath passes above the electroscope.

Exp. 98. Rub an ebonite rod with flannel. Bring it gradually near the disc of an electroscope. Notice that the leaves diverge slightly when the rod is some distance off, and gradually become farther apart as it approaches. Withdraw the rod slowly and notice the gradual collapse of the leaves.

Exp. 99. Perform Exp. 98 again, but place a glass plate between the electroscope and the rod. Similar results take place.

The chief lesson we learn from these experiments is that electricity acts across a dielectric (non-conductor). In Exp. 98 the dielectric was air; in Exp. 99 glass, or rather air and glass. The results are explained by saying that when an electrified body is brought near, but not in contact with, an insulated conductor, the electrified body acts upon the conductor: (*a*) if uncharged *it will attract opposite electricity to the side of the conductor near to it, and repel the same kind to the side remote from it*; or (*b*) if already charged *it will disturb that charge.*

Such electrical action is called **Induction.** The electrified body which produces the action is called the *inducing body*; the electricity produced by the action is called *induced electricity.*

Exp. 100. To determine what actually happens on an insulated conductor when an electrified body is brought near. Connect an

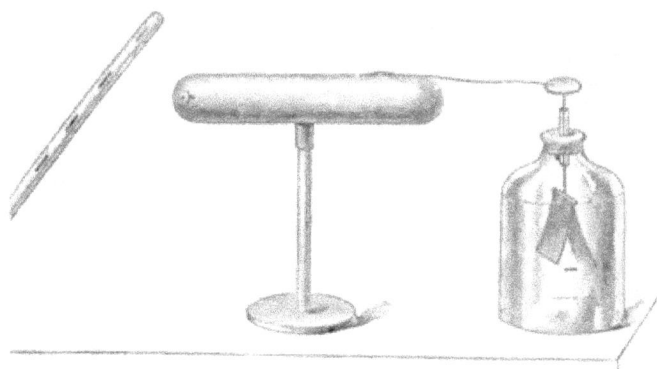

FIG. 79.

electroscope, by means of a copper wire, with an insulated conductor (fig. 79). These form for electrical purposes *one* conductor. Bring near a vulcanite rod rubbed with flannel. The leaves diverge. Remove the wire by means of a non-conductor, e.g. a glass rod. The leaves remain divergent. Bring the rubbed vulcanite rod near the electroscope. Notice that the leaves diverge farther ; therefore the electrification of the leaves is negative, i.e. the repelled electricity is of the same kind as that on the electrified body.

This fact may also be proved by the following experiment.

Exp. 101. Take a thoroughly well-insulated cylindrical conductor with rounded ends. Bring a negatively electrified rod near one end and touch the other end with a proof-plane. Prove that the charge on the proof-plane is negative by means of a negatively charged electroscope.

Exp. 102, to show that a positive charge is attracted to the end of the conductor next the inducing rod. Touch this end with a proof-plane and show that there is a greater divergence of the gold leaves in a positively charged electroscope when the proof-plane is brought near.

We have thus *proved* that a negatively charged rod has acted on the conductor, attracting positive to the end near the rod and repelling negative.

An experiment similar to Exp. 100 may be done as follows :—

Exp. 103. Place two insulated brass balls in contact (two spherical bedstead knobs, two or three inches in diameter, supported on ebonite rods are excellent, and may be made at a cost of a few pence). Bring near a positively electrified rod. As shown in fig. 80, negative electricity is attracted near the rod, and positive repelled. To prove this, remove the ball remote from the rod, taking care to touch the base of the stem only. Remove the rod, and test the kinds of electrification with a gold-leaf electroscope.

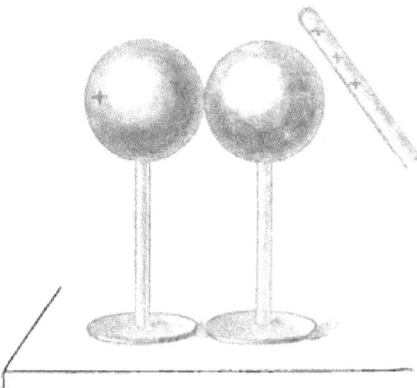

FIG. 80.

Eggs resting on wineglasses may be used for this experiment.

The inductive action takes place through a series of conductors. This may be shown as follows—

Exp. 104. Arrange insulated conductors as shown in fig. 81. Bring up a positively charged rod.

The positive charge on the rod acting on the conductor nearest to it attracts negative electricity and repels positive, which positive, acting on the second conductor, attracts negative and repels positive, and so on. The inductive influence, however, becomes feebler as the distance from the inducing body increases.

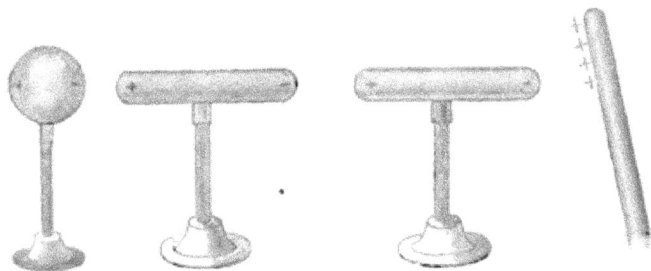

FIG. 81.

Prove the truth of these statements by means of a proof-plane and a gold-leaf electroscope, as described in Exs. 101, 102.

Effect of connecting a Conductor with the Earth.—We have previously learnt that a conductor must be insulated to enable it to exhibit any signs of electrification. Let us now study this subject still further. We have found from the experiments just described that if a charged rod be

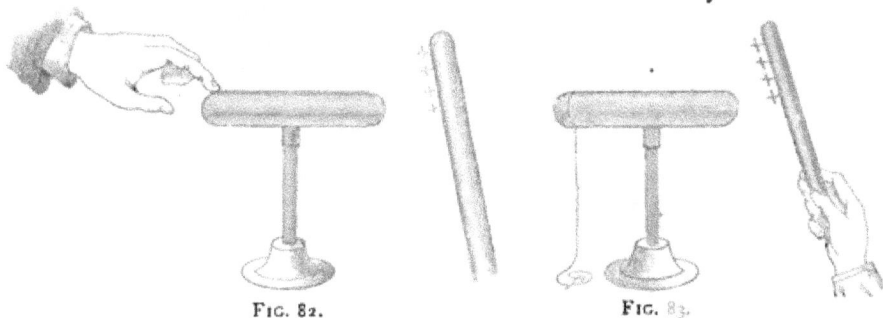

FIG. 82.　　　FIG. 83.

brought near an insulated conductor the opposite electricity is attracted, and similar electricity repelled. If, however, the conductor be placed in connection with the earth by means of a wire, cotton thread, the hand, or in fact any conducting body, as shown in figs. 82, 83, and a positively electrified rod be brought near, negative electricity will be attracted to the end of the conductor near the rod, and positive repelled, which escapes through the conducting body to the earth.

Exp. 105. Having done this, remove the hand or wire, then the rod, and test by means of a proof-plane and a gold-leaf electroscope that there remains a negative charge only : the positive electricity, therefore, has disappeared.

Remember that any part of the conductor may be touched by the conducting body.

Free and Bound Electricity.—Why is this? Why does not the electricity near the inducing body escape to the earth when that part is touched?

The electrified body, after acting inductively on the conductor, holds, as it were, the opposite kind captive by its attracting power. This electricity is then said to be *bound*. The like kind, however, is free to escape if a conducting body be brought in contact with the insulated conductor, and for this reason the electricity is called *free*. Thus in fig. 80 the negative is bound, the positive is free.

Further information respecting 'free' and 'bound' electricity will be given in Chap. XIII.

Inductive Process of Charging a Gold-leaf Electroscope.— The common method of charging a gold-leaf electroscope is by

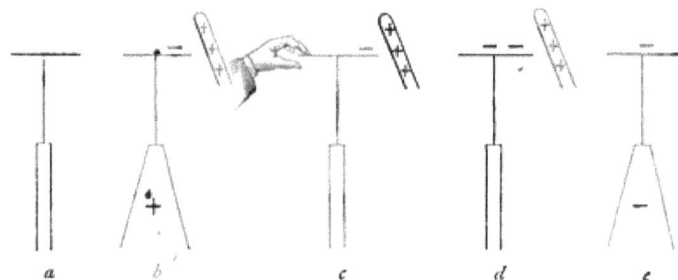

FIG. 84.

the process of induction. The action is perfectly simple, and will be readily understood by aid of fig. 84.

a represents an electroscope in its neutral or uncharged state.

b shows the action when a positively charged rod is brought near the disc ; negative electricity is attracted, and positive repelled into the leaves, **which** therefore diverge with positive electricity.

c. When the disc **is** touched with the hand the leaves collapse, owing to the escape of the free positive electricity.

d. The hand is then removed, the negative electricity being held bound by the positive on the rod.

e. **The rod is** finally withdrawn, **and** the negative electricity, becoming free, spreads over the **whole** conductor—disc, wire, and leaves.

Remember to remove **the** *hand* **first,** then the *rod*.

Exp. 106. Test the charge in the leaves obtained in the above process, by **bringing up a** negatively electrified rod. Notice greater divergence ; therefore the leaves are negatively **electrified.**

Exp. 107. Charge the electroscope positively **in a** similar manner by using a negatively electrified **rod.**

Notice that, if a charge of a particular kind **of** electrification is required in the leaves, a rod oppositely **electrified** is **used.**

Exercise XVII.

1. An egg-shell **is placed on a table, and a** glass **rod** which has been rubbed with silk **is brought near the shell ;** the shell rolls after the rod. Describe the condition of **the rod and the shell** during the motion of the latter.

2. If an electrified **piece of metal is made to** touch a gold-leaf electroscope, **the** leaves separate, and, on **taking the** metal away, they remain separate. **But if the electrified** metal **is only brought** *near* the electroscope, and then **taken away, the** leaves separate **when** the **metal** is near, **but** fall together when **it is** taken away. Why **is there** a lasting effect on the gold leaves in one case, and only a temporary effect in the other ?

3. A lath **six** feet long is supported **at** its centre on a dry glass tumbler. Below one end **of** the lath, and at **a** distance of some inches from it, are placed some scraps **of** gold-leaf **or** other light **bodies. A** glass rod electrified by friction **is** brought over the **other end of the** lath without **touching it.** The **fragments** of gold-leaf are immediately attracted. **How** is this attraction **produced ?**

4. **A stout stick of** sealing-wax **is** stuck upright to a piece **of** wood acting as a base ; **into the wax** at the **top** is inserted a needle, and on the needle is fixed **an apple. Near to** the apple, but not in contact with it, is brought a rod of glass which **has** been rubbed by silk. What is the condition of the apple while the **rod** remains near it ? What occurs when the apple is touched for a moment ? What, **finally,** occurs when the rubbed glass is withdrawn ?

5. Describe and explain the methods of charging the common gold-leaf electroscope.

Induction precedes Attraction.—M, fig. 85, is an insulated conductor charged with positive electricity.   N is a light ball suspended by a silk thread.

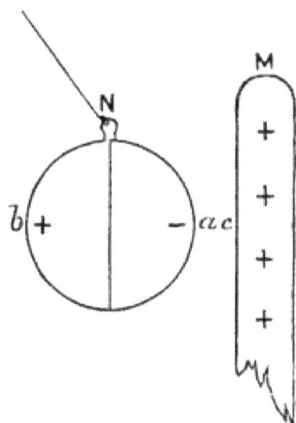

The positive charge on M induces a negative charge on the side of N nearest M and repels a positive charge to the farther side.

The positive charge on M, therefore, *attracts* the negatively charged portion of N, and *repels* the positively charged portion, but c is nearer a than it is to b ; and, since the force of attraction or repulsion varies inversely as the square of the distance (see p. 121), the attraction between a and c is greater than the repulsion between b and c ;

FIG. 85.

therefore the body N, as a whole, is attracted to M.

Faraday's 'Ice-pail' Experiments are of great importance in order to more thoroughly comprehend the principles of induction, and to prove that the charge derived from induction is equal in amount to that obtained from actual contact of the electrified body.

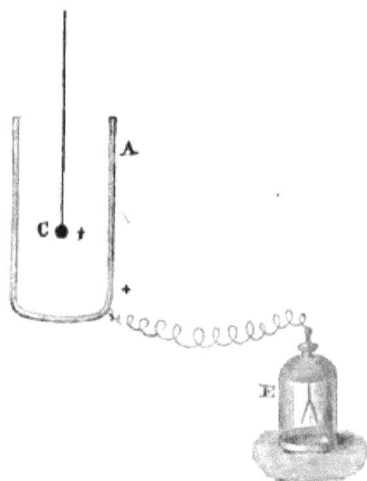

FIG. 86.

Exp. 108. Carefully insulate a metallic vessel A about eight inches deep and four inches in diameter (a tin canister answers excellently).   Connect the outside by means of a wire with the disc of a gold-leaf electroscope, E. Suspend a positively electrified metal ball, C, by an absolutely dry silk ribbon, two or three feet long, and lower it into the vessel.

Immediately the electrified ball enters the pail the leaves of the electroscope begin to diverge. They reach their greatest divergence when the ball is a short distance from the top of the vessel, nor do they alter when the ball is allowed to descend lower. If, now, the ball be removed the leaves at once collapse. The action is perfectly simple. The positively charged ball induces a negative charge on the inner side of the vessel, and repels a positive charge, which, entering the leaves, causes them to diverge. On withdrawing the ball induction ceases, and the two induced charges neutralise each other.

Exp. 109. Repeat the above experiment, but when the ball is quite inside the vessel, allow it to come in contact with the vessel. Notice that the leaves remain divergent to the same extent as before contact. Remove the ball and bring it near a pith-ball pendulum. It will be found to have lost all its electricity. The charge has, therefore, been given up to the vessel ; and, since the leaves of the electroscope are not affected, the positive charge on the ball has exactly neutralised the induced negative charge. We also conclude that the positive induced charge is exactly equal to the positive charge on the inducing body, for we know that the positive charge induced on an insulated conductor is always equal to the negative charge induced on the same conductor.

Exp. 110. Prove that the electrification of the leaves in Exp. 109 is positive by bringing near the electroscope a positively charged rod.

Exp. 111. Repeat Exp. 108, but while the ball hangs in the vessel touch the vessel or the electroscope with the finger. The leaves collapse, owing to the removal of the positive induced charge. Remove the hand, then withdraw the ball ; the leaves diverge.

Exp. 112. Prove that this electrification is negative, by bringing up a negatively charged rod.

Exp. 113. Repeat the first part of Exp. 111 ; after removing the positive electricity touch the vessel with the positively charged ball. On withdrawing the ball, the leaves show no evidence of electrification, as the induced negative charge is neutralised by the inducing positive charge.

Inductive Capacity.—The inductive action is not so simple as it appears at first sight. We have thus far assumed that electricity acts at a distance, and have disregarded the medium across which it acts.

Faraday, by a series of remarkable experiments, proved that when a charged body acts inductively on another body, the medium between them performs an important office. According to Faraday's inductive theory it is assumed that the molecules of every dielectric act as conductors, and are separated from each other by a non-conducting medium. These molecules are acted on in turn, one half becoming positive and the other negative. Suppose, for example, that the medium

FIG. 87.

is air, and let us further suppose that we can represent the molecules by small circles. Fig. 87 will assist the student to grasp the action. Let A be a positively charged conductor, and B the body acted on by induction. First, the layer of molecules of the air next to A is inductively acted on, the nearer half of each molecule becoming negatively and the remote half positively electrified. This positive electricity acts similarly on the next layer, and so on, until, finally, the layer of molecules next to B acts upon it, as shown in the figure.

The action, therefore, is *transmitted* from A to B. Faraday called this action *dielectric polarisation*.

The power of transmitting induction varies with different bodies. Across substances such as glass, shellac, or sulphur, the effect produced by the electrified body is greater than across air, the distance remaining the same.

The power a body has of allowing induction to take place across it is called its *inductive capacity*.

The Electrophorus.—This simple instrument was invented by Volta in 1775, for obtaining a series of charges of electricity from a single charge. It consists of (1) the 'generating plate' D, fig. 88—a flat round cake of resin, sealing-wax, shellac, or vulcanite [1]

---

[1] The surface of vulcanite should occasionally be washed, first with ammonia and then with paraffin oil.

(the last is excellent) contained within a metal dish or resting upon a metal plate ; (2) this metallic plate or dish is called the 'sole;' (3) an upper disc—the 'cover' or 'collecting plate'

Fig. 88.

A, of slightly smaller diameter, consisting of a conducting body e.g. metal, or wood coated with tinfoil, to the centre of which an insulating handle is fixed.

Method and Theory of Charging the Electrophorus.—
Exp. 114. Warm the generating plate until it is perfectly dry. (1) Rub or strike the plate with warm flannel or fur (a fox's brush or catskin is very useful for this purpose). This generates negative electricity on the upper surface of the disc, which, acting inductively

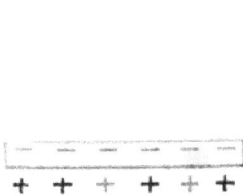

Fig. 89.

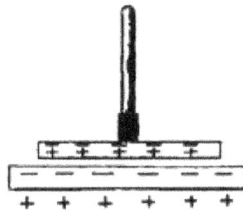

Fig. 90.

through the plate, 'binds' positive on the top of the sole and repels negative, which escapes to the earth (fig. 89). By the action of the 'bound' positive charge the tendency of the negative charge on the plate to escape is diminished.

(2) Place the 'cover,' by means of the insulating handle, on the generating plate. The two discs touch at certain points only, so that there is a thin layer of air between them. We have therefore a negatively charged body separated from a conductor. The consequence is that induction takes place, a positive charge being attracted to the lower surface of the disc and a negative charge repelled to the upper surface (fig. 90).

(3) Touch the 'cover' with the finger. The free negative electricity escapes through the body to the earth ; or we may regard

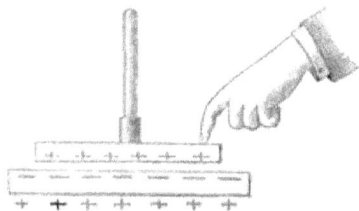

FIG. 91.

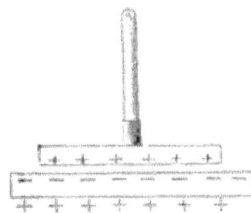

FIG. 92.

it as being neutralised by positive electricity flowing from the earth through the body and hand (fig. 91).

(4) Remove the finger. The positive charge on the 'cover' is bound by the negatively charged plate (fig. 92).

(5) The 'cover' is then lifted by the handle : the positive electricity, being no longer 'bound,' distributes itself over the conductor

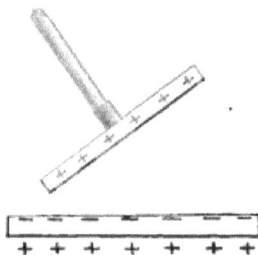

FIG. 93.

(fig. 93). There is a charge sufficiently strong to yield sparks when the knuckle is presented to it.

If the generating plate be perfectly dry 2, 3, 4, 5 may be repeated many times without again exciting it.

We must not be satisfied to *mention* these facts, but we must *prove* that they are true.

Exp. 115, to prove that the generating plate is negatively charged (see 1, Exp. 114).

Charge a pith-ball pendulum or gold-leaf electroscope with negative electricity. Bring the rubbed disc near it. Notice repulsion or further divergence. The electrification of the plate is, therefore, negative.

Exp. 116, to prove that negative electricity escapes from the sole (see 1).

Place the generating plate on the disc of a gold-leaf electroscope. Beat the plate with a fox's brush. The leaves gradually diverge. Remove the plate by two insulating rods. Test the charge in the leaves by means of a negatively charged rod. There is further divergence.

Exp. 117, to prove that the upper surface of the cover is negatively electrified (see 2).

Attach a wire to the disc of a gold-leaf electroscope. Place the cover on the excited generating plate. Use a vulcanite rod to place the other end of the wire on the cover. The electricity passes along the wire and the leaves diverge. Test the kind; they will be found to be negatively charged. Care must be taken with this experiment, or the divergence may be so great as to fracture the leaves.

Exp. 118, to prove that the cover is positively electrified (see 4 and 5).

Charge a pith-ball pendulum positively. Bring the cover near. Repulsion takes place.

To obviate the necessity of touching the cover with the finger a brass pin is sometimes placed in the centre of the lower disc, one end of which is in contact with the sole and the other just reaches the upper surface. Thus the pin touches the collecting plate each time, and allows the free electricity to escape to the earth. Instead of a pin a strip of tinfoil sometimes passes from the sole to the upper surface of the generating plate, so that the cover touches it each time.

Exp. 119. A very interesting and instructive experiment can be done by taking a small tin cover (e.g. the cover of a mustard tin), and filling it with molten sealing-wax. This forms the generating plate. Fasten a well-worn penny to a stick of sealing-wax. This forms the collecting plate. A balanced lath can be attracted by the penny piece if charged as described in Exp. 114.

Exp. 120. Support a metal tray on shellac-varnished tumblers. Electrify hot brown paper with a clothes brush and lay it immediately on the tray. Touch the tray with a proof-plane and show that it is negatively charged. Touch the tray with the finger. Lift the paper and show that the tray is positively charged.

EXERCISE XVIII.

1. An insulated conductor, A, is brought near to the cap of a gold-leaf electroscope which has been charged positively. State and explain what will happen (1) if A is unelectrified, (2) if it is charged positively, (3) if it is charged negatively.

2. A brass rod is supported horizontally by a dry glass stem, and a large strongly electrified metal ball is brought near one end of the rod (but not near enough for a spark to pass). The rod is then touched for an instant by the end of an earth-connected wire, and afterwards the ball is removed. Will it make any difference in the final electrical state of the brass rod whether the wire touches it at the end nearest the ball, at the end farthest from the ball, or at the middle? Give reasons for your answer.

3. Supposing you were required to develop induced electricity, and to prove its existence, how would you do it?

4. A stick of sealing-wax, after being rubbed with flannel, is found to be negatively electrified. How, by means of it, would you charge a proof-plane with positive electricity?

5. Two insulated metal spheres are placed in contact with each other. A positively charged glass rod is brought near one of the spheres, and while it is there the other sphere is removed. The glass rod is now taken away. On bringing the spheres near together again a spark passes between them. Give the reason of this.

6. I grind sulphur in a mortar, and thus electrify it. I place some of the electrified powder on the plate of a gold-leaf electroscope. The leaves diverge; why? since the sulphur is an insulator, and does not part with its electricity. I remove the powder by means of an insulator: what will occur, and why will it occur?

7. Describe the electrophorus and the mode of charging it.

8. Give the theory of the action of the electrophorus.

9. A little pith ball rests on a brass plate provided with a glass handle. The two are placed on a cake of resin which has been rubbed with a cat's skin. When the plate is touched with the finger, and then lifted by the handle, the pith ball jumps off the plate. Why?

10. I fasten a stick of sealing-wax to a penny or a halfcrown, and I whisk a piece of vulcanised india-rubber with a fox's brush. Holding the sealing-wax as a handle, I lay the penny or halfcrown flat on the india-rubber. What is the condition of the coin? I touch the coin: what happens? I lift it by the handle: what is its condition?

11. Explain Exp. 120.

12. Explain, by means of a diagram, how induction precedes attraction.

CHAPTER X.

DISTRIBUTION.

Seat of Charge.—Exp. 121 (commonly called Biot's or Caven-
dish's experiment). Charge an insulated metal ball (four inches
in diameter) by means of an electrophorus, or better, by means of
an electrical machine (to be described hereafter). Place two
hemispherical envelopes furnished with glass handles on the out-

FIG. 94.

side (fig. 94). After contact with the sphere, carefully remove the
hemispheres, and bring each of them near an uncharged gold-leaf
electroscope ; in each case the leaves diverge : now bring the ball
near the electroscope ; the leaves do not diverge, although the
ball originally received the charge.

Why is this? When the covers are placed over the ball they
form the outside of the conductor ; and, as they have been proved
to be charged, while the ball has lost its charge, the electricity
must have passed from the surface of the ball to that of the covers

Generally, we find that a charge of electricity imparted to an insulated conductor instantly distributes itself over its *outer surface*. Even if the conductor be hollow and the charge be imparted to its interior, the result is precisely the same—it immediately flows to and over the surface.

The seat of charge of statical (or frictional) electricity is therefore the outer surface of conductors.

Many experiments further prove this law.

Exp. 122, to prove that two conductors of the same size, one solid and the other hollow, receive equal charges.

Take two small boxes of equal size (metallic match-boxes answer very well), one empty, and the other full of iron filings. Place them in contact on an insulator. Charge them, and remove one some distance from the other by means of a non-conductor. Touch each with a proof-plane and notice that an equal divergence of the gold leaves of an electroscope is produced in each case.

Exp. 123. Take a *hollow* insulated metal ball having an aperture at the top of about 1-1½ inches in diameter (fig. 95), and charge it with positive electricity. Apply a proof-plane (*c*) to the interior, and bring it in contact with an uncharged electroscope ; no action takes place ; there is therefore no electrification on the inside. Now touch the outside with the proof-plane and present it to the electroscope. The leaves diverge at once. The electricity is therefore found on the outer surface only.

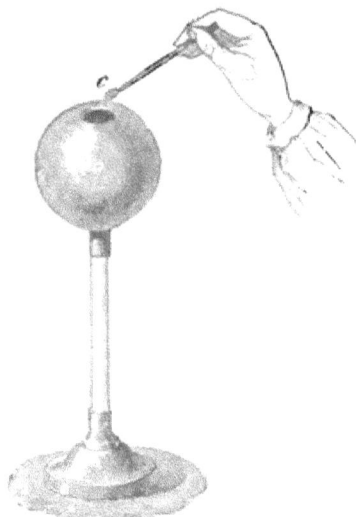

FIG. 95.

Exp. 124. Electrify an insulated metallic vessel or wire gauze cylinder (fig. 96). Touch the inside with a proof-plane and bring it in contact with a gold-leaf electroscope. There is no action. Test the outside ; the leaves diverge. If, however, when touching the interior, part of the proof-plane is above the edge of the vessel, a small charge will be found upon the proof-plane. In this case the latter formed part of the outside of the vessel.

Exp. 125 (known as Faraday's 'butterfly net' experiment). Make a conical muslin or gauze bag (of the shape commonly used for catching butterflies). Fasten this to a brass ring which is fitted on an insulating support (fig. 97). To the apex of the bag attach two silk threads, by means of which the bag can be drawn inside out. Electrify the bag. (*a*) Test the inside by means of a proof-

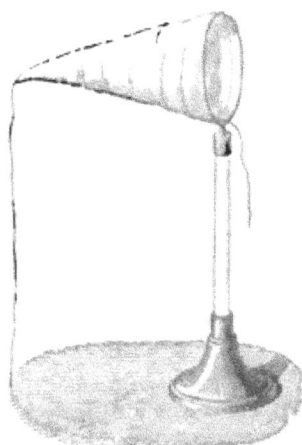

FIG. 96.   FIG. 97.

plane and electroscope. No action. (*b*) Touch the outside with the proof-plane. The leaves diverge when the proof-plane is brought in contact with the electroscope. Touch the proof-plane and electroscope so as to discharge them. (*c*) Turn the bag inside out by pulling the silk thread. Touch the inside (the former outside, where the electricity was found to reside) with the proof-plane. No action. Similarly test the outside (which was formerly the inside). The leaves diverge. The charge has therefore passed to the outside of the bag.

Exp. 126 (devised by Terquem). Insulate a wire cage and suspend inside it a pair of gold leaves or pith balls by cotton threads. Attach the cage, by means of a chain, to the prime conductor of an electrical machine. Set the machine in motion; no divergence of the leaves or pith balls is noticed. The electricity is found on the surface to such an extent that sparks may be drawn from it when the knuckle is brought near.

It may be mentioned as a further illustration of this fact that Faraday had a room built, each side of which measured

12 ft.　He describes it as follows : ' A slight cubical wooden frame was constructed, and copper wire passed along and across it in various directions, so as to make the sides a large net-work, and then all was covered in with paper, placed in close connection with the wires, and supplied in every direction with bands of tinfoil, that the whole might be brought into good metallic communication, and rendered a free conductor in every part.　This chamber was insulated in the lecture-room of the Royal Institution. . . . I went into the cube and lived in it, and used lighted candles, electrometers, and all other tests of electrical states.　I could not find the least influence upon them, or indication of anything particular given by them, though all the time the outside of the cube was powerfully charged, and large sparks and brushes were darting off from every part of its outer surface.'[1]

Applications.—Delicate apparatus may therefore be en-veloped by wire, or even cotton-gauze, to screen them from the influence of electrified bodies during experimental work. It has also been suggested to cover buildings with wire to pro-tect them from lightning.

Exceptions.—There are, however, two exceptions to the law that electricity resides on the outer surface of bodies.

1. In voltaic electricity, the current flows *through* conductors.

2. When we place an insulated electrified body inside a hollow conductor, inductive action is set up, and we have a charge on the interior surface.

Exp. 127.　Insulate a metallic vessel.　Suspend a metal ball by a silk ribbon and charge it positively.　Lower the ball into the vessel.　By means of a proof-plane and a gold-leaf electroscope, show that a negative charge can be obtained from the inside of the vessel.

Electric Density.—The three following experiments are by no means sufficiently accurate as exact quantitative measure-ments, but they are suitable for our purpose.

Exp. 128.　Charge a thoroughly well-insulated metal sphere with positive electricity.

---

[1] *Experimental Researches*, Nos. 1173, 1174.

(*a*) Touch some point on the surface with a proof-plane and bring it in contact with an uncharged gold-leaf electroscope. Notice the amount of divergence.

(*b*) Discharge the proof-plane and the electroscope.

(*c*) Touch a different point on the sphere with the proof-plane and touch the electroscope as before. Notice that there is an equal divergence of the leaves.

Exp. 129. Electrify an insulated pear-shaped conductor (fig. 98).

FIG. 98.

(*a*) Touch the larger end of the conductor with a proof-plane and bring it in contact with an electroscope. Notice the amount of divergence of the leaves.

(*b*) Discharge the proof-plane and the electroscope.

(*c*) Touch the pointed end, *a*. Notice that on contact with the electroscope the divergence of the leaves is greater. In this case, therefore, the charge is not uniformly distributed.

Exp. 130. Electrify a round sheet of tin placed on a varnished glass tumbler : (*a*) test the middle of the plate with a proof-plane and a gold-leaf electroscope. Notice divergence : (*b*), test the edge of the plate and observe that the divergence is greater.

Thus we conclude that the electricity is equally distributed over the surface of the sphere, but that it is unequally distributed on the pear-shaped conductor and on the disc. In other words, the electricity is of *equal density* on the sphere, but of *unequal density* on the pear-shaped conductor and on the disc.

Electric density is defined as the quantity of electricity on a body per unit area.

Relation between electric density and area.—Exp. 131. Attach, by means of gummed paper, a sheet of tinfoil to a

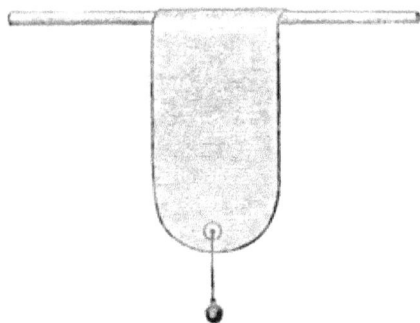

FIG. 99.

varnished glass rod. Cut the other end so that the corners are removed, as shown in fig. 99. Gum a circular piece of paper near the bottom, and through the centre fasten a silk thread, to the end of which a small metal ball is hung, sufficiently heavy to ensure the tinfoil being tightly wrapped round the rod. Thoroughly dry the rod and charge the open sheet of tinfoil. Touch a point near the bottom with a proof-plane, and bring it in contact with a gold-leaf

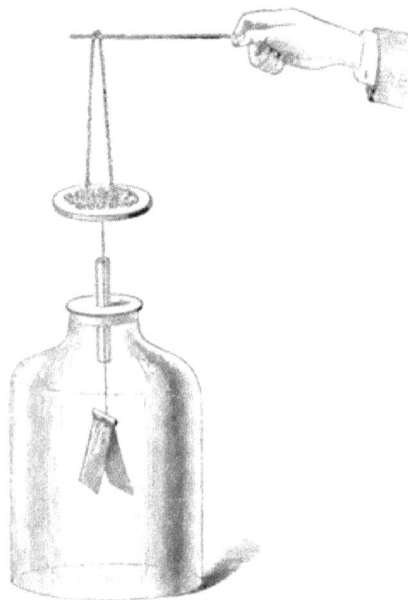

FIG. 100.

electroscope. Notice that the leaves diverge. Roll up the sheet on the rod; test the same place with the proof-plane and the electroscope. The leaves diverge more widely. This is due to the fact that *the density increases as the surface becomes smaller.*

Exp. 132. Place a chain on the disc of an electroscope, fig. 100. Charge the latter by induction, so that the leaves are widely divergent.

1. Raise the chain with an ebonite rod; the leaves slowly collapse as the area becomes larger.

2. Lower the chain; as the surface diminishes, the density increases, and the leaves again diverge.

[The last two experiments also illustrate the law that electrification resides on the outside of conductors.]

Generally, if S represents the area of a conductor,

Q   ,   „   quantity of electricity,

and *d*   „   „   density,

then $d = \dfrac{Q}{S}$

If, therefore, we double the quantity of electricity on any given surface, we double the density.

If we halve the quantity, we halve the density.

If we increase the surface, the quantity remaining the same, we diminish the density, and *vice versâ*.

Quantity of Electricity.—As we have had occasion to use the expression *quantity of electricity*, it becomes necessary to explain its meaning.

The electrification or charge of an insulated conductor is a *measurable quantity*, which can be ascertained by means of electrometers. For the present it must be remembered that when we speak of a quantity of electricity we mean a charge of a definite number of units. *By a charge of one unit is meant that charge on a very small body, which, if placed at a distance of one centimetre from an equal and similar charge, repels it with a force of one dyne.*[1] *The medium between the two charges is assumed to be air.*

EXERCISE XIX.

1. An insulated sphere of 4 centimetres radius has a charge of 128 units. What is the surface density?

Generally $d = \dfrac{Q}{S}$

Now the surface (area) of a sphere = (diameter)² $\times \dfrac{22}{7}$

$\therefore$ surface $= 8^2 \times \dfrac{22}{7}$

whence $d = \dfrac{128 \times 7}{64 \times 22}$

$= .63$ nearly.

[1] In scientific works the fundamental units are:—

The centimetre (·3937 inch) as the unit of length.
The gramme (15·432 grains)   „   „   mass.
The second   „   „   time.

Other units are derived from these—e.g., the *unit of force* is that force which,

2. What is meant by electric density?

An insulated sphere of 5 centimetres radius has a charge of 200 units. What is the density? *Ans.* : ·63 nearly.

3. A sphere has a radius of 10 centimetres. It is insulated and then charged until its surface density has a value of 14. What quantity of electricity was used? *Ans.* : 17,600 units.

4. What quantity of electricity is required to charge a sphere 14 centimetres in diameter so that its surface density has a value of 5?
Ans. : 3,080 units.

5. A sphere is charged with 3,168 units of electricity, and its surface density is 7. What is the radius? *Ans.* : 6 centimetres.

Surface density on differently shaped conductors.—On differently shaped insulated conductors the distribution varies.

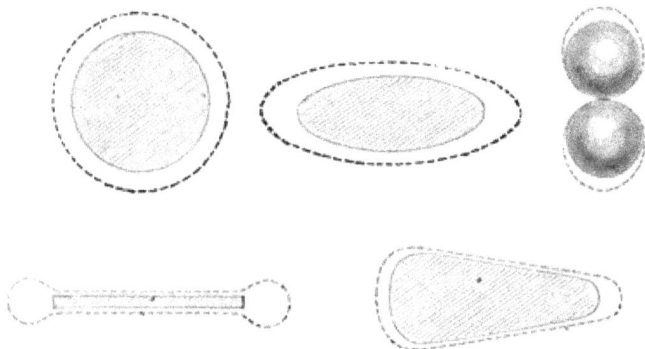

FIG. 101.

In fig. 101 we may imagine that the density of the electricity varies according to the distance of the dotted lines from the conductors.

On the sphere the density is uniform.

On the ellipsoid the density is greatest at the ends.

On the pear-shaped conductor the density is greatest at the pointed end, less on the broad, and least on the flat part.

On the disc the density is greatest at the edge, uniform and small on the flat sides.

On a cube the density is greatest at the corners, less at the edges, and least on the sides.

Coulomb performed many quantitative experiments on distribution. He found that on a cylinder, having round ends,

acting for one second on a mass of one gramme, gives it a velocity of one centimetre per second. This is called a **dyne**.

30 inches long, and 2 inches in diameter, the density was 1 at the middle; 1·25, two inches from the ends; 1·8, one inch from the ends; and 2·3 at the ends.

Remember that the distribution, as shown in fig. 101 is correct only when the conductors are remote from the influence of other electrified bodies.

Action of points.—We will now consider more closely the charge on an insulated conducting ellipsoid (fig. 102). The law respecting the density at the ends of the axes is given in the following proportion :—

Density at A *or* B : density at C *or* D :: A B : C D.

If A B is twice C D, then the density at A is twice that at C.

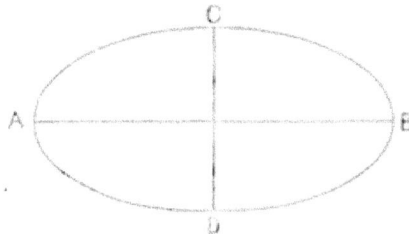

FIG. 102.

If A B is ten times C D, then the density at A is ten times that at C.

Thus by extending A B, we have eventually a long, finely pointed body, at the end of which the density may be exceedingly great—e.g., if A B be one hundred times C D, then the density at the end A or B is one hundred times that at C or D.

In fact, as bodies become pointed, the electricity accumulates more and more at the pointed end, until the density becomes so great that the electricity is discharged. The particles of air surrounding the point become electrified, and are, therefore, repelled ; other particles rush in to take their place, which in turn are electrified and repelled. A current of air, called the 'electric wind,' is thus produced (see pp. 119, 120).

Remember that points on conductors **cause a** *continual* **loss of** *electricity* ; they must, therefore, be carefully avoided in all apparatus where they are not essential.

Exp. 133. Electrify a shellac rod. Pass the point of a sharp needle rapidly from one end of the rod to the other two or three times ; repeat this on the other side. By means of an uncharged electroscope, show that the rod is completely discharged. Further experiments on the 'action of points' will be found on pp. 118–120

EXERCISE XX.

1. If you were given a negatively electrified stick of sealing-wax and two metal balls mounted on insulated supports, how would you, with the apparatus, charge the balls with opposite kinds of electricity? How could you afterwards find out whether you had charged the balls as you intended, and whether their charges were equal or unequal?

2. Give experiments which prove that statical electricity resides on the surface of conductors.

3. A deep metal pot positively electrified stands on a glass stem. A metal ball, hung by a silk thread, is put in contact with a gold-leaf electroscope after being made to touch—

(*a*) first the inside and then the outside of the pot ;

(*b*) first the outside and then the inside of the pot.

State and explain the effect on the electroscope in each case.

4. To protect a gold-leaf electroscope from being acted on when an electrical machine is at work near it, it is sufficient to cover the electroscope with a thin cotton cloth. How is this?

5. A pewter pot is insulated and electrified. If you touch it at different parts with a penny stuck to the end of a rod of sealing-wax, what part of the pot will give the greatest quantity of electricity to the penny.

6. I have an insulated solid brass ball and an insulated hollow brass ball of equal diameter. I electrify them so that each receives the same number of units of electricity. I test them with a proof plane and a gold-leaf electroscope. Will there be any difference in the indication? If so, what? If not, why not?

7. (*a*) I insulate and charge a square foot of tinfoil, and observe the effect on a gold-leaf electroscope.

(*b*) I make a ball of the tinfoil, insulate and charge it with an equal number of units of electricity as in (*a*). I observe the effect under similar conditions on the electroscope. Will there be any difference in these effects? Explain.

Electric tension.—That which produces a tendency in electricity to escape from a conductor on which it is accumulated is sometimes called *electric tension*—although happily the term is falling into disuse. It has been well defined as the *mechanical stress* [1] *across a dielectric.* Tension, therefore, is of the nature of a force, and must not be confounded with the term density.

Although tension varies with the density (the law being that electric tension is proportional to the square of the electric

---

[1] A *strain* is an alteration in form or volume by means of a *stress.* A *stress* is a force, or indeed any agency, which produces a *strain.*

density), we must remember that the notion conveyed by the two terms is essentially different.

Redistribution and Subdivision of Charges.—If we have a charged insulated conductor unaffected by other electrified bodies, and any portion of that charge be removed, the remaining electrification will redistribute itself over the surface in a manner similar to the distribution of the original charge. This follows from the fact that electricity in equilibrium can only be distributed in one way upon a given conductor.[1]

Suppose, for example, that we have several insulated metal spheres of equal size, one of them being charged with a certain quantity of electricity, and the others uncharged. If one of the latter be brought in contact with the charged sphere, it will

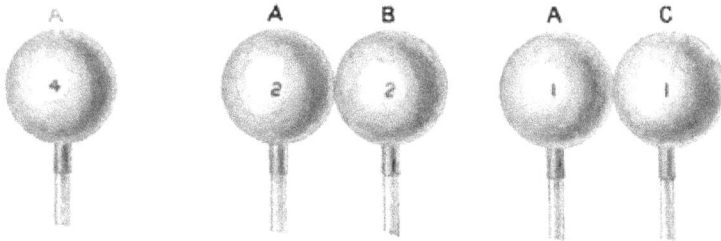

FIG. 103.

receive half the charge ; if on separating the spheres, another be brought in contact with either, it will receive half *its* charge, i.e. one quarter of the original charge. The distribution in each case will be uniform.

Thus let A (fig. 103) receive four units of electricity.

Put A in contact with B ; each then has two units. Re-

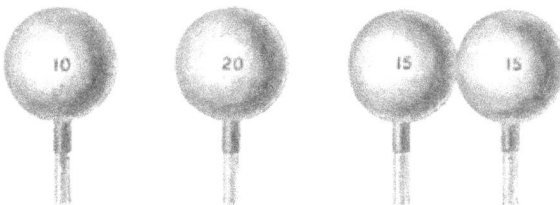

FIG. 104.

move B, and put A in contact with C. Each of these now has half of two units, i.e. one unit.

[1] For proof of this statement, a mathematical treatise must be consulted.

Again, let two equal insulated spheres be charged, one with ten units of positive electricity, and the other with twenty units of positive electricity, and placed in contact; each now has half the sum of the two charges, *i.e.* fifteen units (fig. 104).

A similar result occurs if one of them has twenty units of

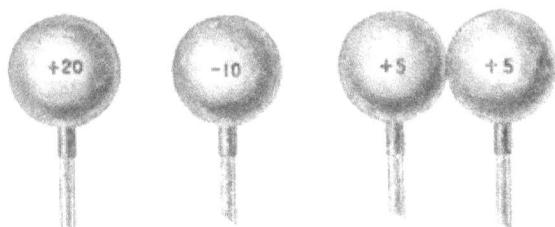

FIG. 105.

positive (+) electricity, and the other ten units of negative (−) electricity (fig. 105).

On contact each has $\dfrac{+\ 20\ -\ 10}{2} = \dfrac{10}{2}$

= 5 units of positive electricity.

The consideration of the subdivision of charges on two unequal spheres is given on p. 107.

CHAPTER XI.

POTENTIAL AND CAPACITY.

Potential.—We must now discuss, in a very simple manner, the meaning and use of the term *potential* with reference to electricity. The idea of electric potential involves a mathematical conception which cannot be adequately treated of in an elementary work ; the term is, however, of such great importance and value that it is necessary to gain some knowledge respecting it.

First notions :—(*a*) When a positively charged conductor is connected with the earth, electrification is transferred *from* that body to the earth.

(*b*) When a negatively charged conductor is connected with the earth, electrification is transferred *to* that body from the earth.

(*c*) If two insulated charged conductors be placed in contact, or in metallic connection, electrification may or may not pass from one to the other. Now, whether electrification passes from one of the conductors to the other or not, depends upon the potentials of the conductors.

Thus in (*a*) the body is said to be at a higher potential than the earth ; in (*b*) the earth is said to be at a higher potential than the body ; in (*c*) if electricity passes from one body to the other, the body *from* which the electricity flows is said to be at a higher potential than the one *to* which it flows.

Definition.—The potential of a conductor may therefore be defined as the electrical condition of that conductor which determines the direction of the transfer of electricity.

When the bodies concerned are removed from the influence of other bodies, we may speak of a positively electrified conductor as one electrified to positive potential, and a negatively electrified one as being at negative potential. On the other hand, two insulated conductors in contact under the inductive influence of a charged body may be differently charged though at the same potential.

If no electricity flows, the bodies are at the same potential.

The student will, therefore, see that the term potential as here used is relative, i.e. we compare the potential of one body with that of another. It is convenient to have a standard of reference whose potential is considered to be zero. The earth is usually assumed to be at zero potential for experimental purposes. The potential of a body is the difference between its potential and that of the earth.

Analogies.—Potential is analogous in many respects to temperature and level.

Temperature.—(1) When we say that the temperature of boiling water is 100° C., we mean that it is 100° above a standard point of reference, viz. that of melting ice (0° C.) ; or if we say that the temperature on a certain night was —20° C., we again mean that it is 20° C. below 0° C.

(2) If two bodies at different temperatures are put in contact, or in thermal communication, heat passes from the body at higher temperature to the one at a lower, and will continue to do so until the temperatures are equal.

Level.—(1) If we say that a mountain is 10,000 ft. high, we mean that it is 10,000 ft. above a standard point of reference—the sea-level ; or, if a mine is 1,000 ft. deep, we again mean that it is 1,000 ft. below the same standard point. The sea-level is a convenient point of reference for differences of level.

(2) If two or more vessels containing water at different levels be put in communication at their bases, by means of a pipe or pipes, water will flow from the one at higher level to the one at a lower, until the water is at the same level in all the vessels.

When we speak of the level of water in a pond, or the temperature of a body, the *level* is not the *water* itself ; nor is *temperature, heat*; so when we speak of the potential of a conductor, *potential* is not *electricity*, but merely that state of the conductor which determines the transfer of electricity.

These analogies must not be pushed too far, e.g., (1) with respect to *level* and *potential*, electricity resembles a fluid in so far as it is capable of flowing along conductors, as water flows from higher to lower levels. It must be carefully borne in mind that electricity has no mass.

(2) With respect to *temperature* and *potential*; increase of temperature may cause a solid to assume a liquid or gaseous form, but no such physical change is made in a body by raising its potential.

Measurement of difference of electric potential.—*Difference of level* may be ascertained by means of measuring rods or tapes. *Difference of temperature* may be measured in many ways : one of which is by noting the height of a column of mercury in a thermometer. *Difference of potential* can be measured by any instrument which measures charges of electricity. With reference to this measurement let us consider a gold-leaf electroscope. The student has already learnt that we can only use the electroscope to measure charges *roughly*.

When a gold-leaf electroscope stands upon a table in its ordinary uncharged state, the instrument is at the potential of the earth, i.e. zero potential. When an electrified conductor—e.g. a proof-plane—is brought in contact with the disc of the electroscope, the leaves diverge, because they have acquired a potential different from that of the earth. If the proof-plane is charged positively, the potential of the leaves is higher than the potential of the earth ; if it is negatively charged, the potential of the leaves is lower than that of the earth.

From the preceding considerations in this chapter, the student will understand that every point of a conductor, when a charge is in equilibrium, must be at the same potential ; for if one point differed from another in this respect, there would be a continual flow of electricity from the point at high potential to that at a lower. We can prove this fact by experiment.

Exp. 134. Attach one end of a long fine wire (say 5 or 6 feet of No. 33 B. W. G.[1]) to the disc of a gold-leaf electroscope, and the

[1] Birmingham wire gauge—the recognised method of measuring the thickness of wires.

other to a small proof plane. Charge the pear-shaped conductor (fig. 106) by means of an electrical machine. Touch the flat side of the conductor with the proof plane. Observe that the leaves diverge.

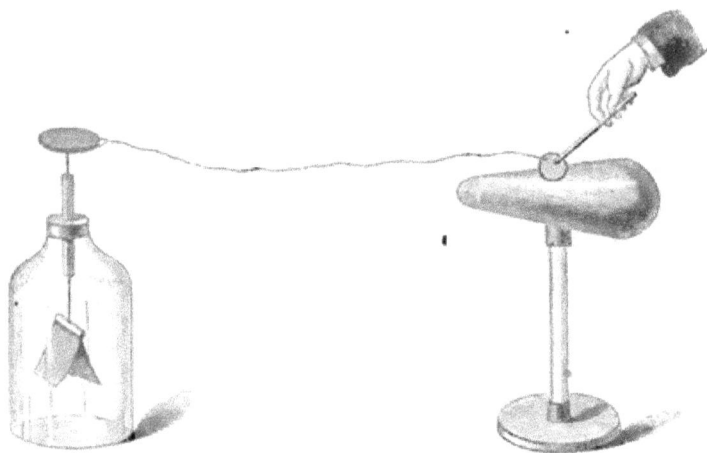

FIG. 106.

Move the proof-plane to the pointed end, and observe that no further divergence takes place.

Thus the divergence of the leaves is independent of the *density* of the electricity on the conductor.

Electrical capacity.—If we take two unequal vessels and fill them with water, the quantity of water poured into each vessel will depend upon the *size* or *capacity* of the vessel. Similarly if we take two insulated conductors of the same shape but of different size, and electrify them, the large one must have a greater charge than the small one to electrify it to the same potential ; i.e., the large one has a greater electrical *capacity* than the small one. Thus, the potential of a conductor depends both upon its charge and its capacity ; in fact, if c = the capacity of a conductor ; Q = the charge or quantity of electricity ; and v = the potential ;

$$\text{then } c = \frac{Q}{v}.$$

Subdivision of charges on spheres of different size.— We are now in a position to understand the subdivision of charge when a small uncharged sphere is put in contact with a larger charged one. They do not share the electrification equally, but the charge depends upon the electrical *capacity* of the spheres.

The capacities of spheres are proportional to their radii.

Thus if a sphere charged with fifteen units of electricity has a radius of one inch, and that of an uncharged sphere a quarter of an inch, when the two are brought in contact (fig. 107) and afterwards separated, the quantity on the large one is twelve units; that on the small one, three units :

for let q = quantity on the large sphere

then $\dfrac{q}{4}$ = „ „ „ small „

$\therefore q + \dfrac{q}{4}$ = whole quantity

$\therefore \dfrac{5}{4} q = 15$

$\therefore q = 12$ and $\frac{1}{4} q = 3.$

FIG. 107.

EXERCISE XXI.

1. Two insulated brass balls touch ; one has a diameter of 3 inches, and the other a diameter of 1 inch. A charge of 24 units of electricity is given to them. They are then separated. How will the charge be distributed ? *Ans.* : 18 and 6 units.

2. If one ball has a diameter of $1\frac{1}{2}$ inch and another a diameter of 6 inches, and while they are in contact they receive a charge of 25 units, how is the charge distributed when they are separated ?
Ans. : 20 and 5 units.

3. A large insulated metal sphere is charged with 12 units of electricity ; another sphere of one-fifth the radius is brought in contact. How is the charge distributed when they are separated ? *Ans.* : 10 and 2 units.

4. Two insulated metal balls are in contact ; one has a radius of 4 inches, and the other a radius of 3 inches. They are charged, and on testing the larger one it is found that it has a charge of 8 units. What was the total charge ? *Ans.* : 14 units.

CHAPTER XII.

ELECTRICAL MACHINES.

AN electrical machine is an instrument by means of which we can obtain greater quantities of electricity than by any of the methods previously described. It consists of two parts—(1) a non-conductor, electrified by the friction of the rubber, for *producing* electricity ; (2) the prime conductor, for *collecting* electricity.

The first electrical machine was invented by Otto von Guericke in 1671. It consisted of a ball of sulphur which was rubbed with the dry hand while it revolved.

Sir Isaac Newton afterwards introduced a glass globe. Shortly afterwards a rude prime conductor was added ; then

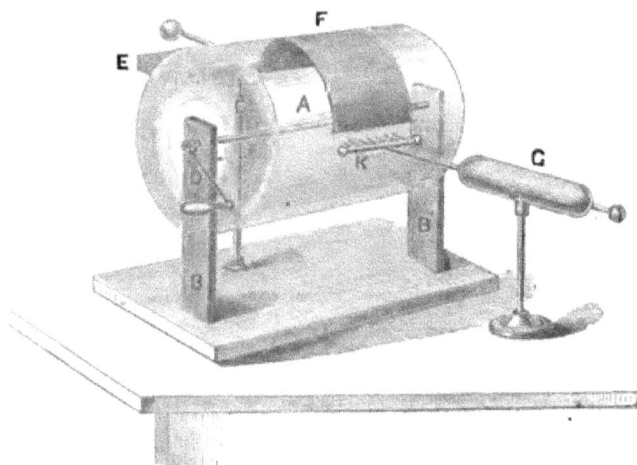

FIG. 108.

a leather cushion was introduced as a rubber ; next the globe made way for a cylinder ; and finally (1760) use was made of a

glass plate. In recent machines plates of vulcanite are sometimes substituted for glass.

The cylinder machine consists of :—

(1) A glass cylinder, A (fig. 108), supported on two wooden
stems, B B', and turning by means of a handle, D, on a horizontal axis.

(2) A rubber, E, consisting of a cushion of leather, stuffed
with horsehair, to which is attached a flap of silk, F, covering
the upper part of the cylinder.

The chief use of this silk flap is to prevent a loss of
electricity. The rubber should be covered with powdered
amalgam [1] and should be pressed upon the cylinder by its
support, C, which may be made of wood or of varnished glass
(the latter is preferable, as it is sometimes convenient to obtain
a negative charge, which we shall learn accumulates on the
rubber ; for this purpose it must be insulated).

(3) The 'prime con-
ductor,' G, consists of a
cylinder, having rounded
ends, made of metal, or of
wood coated smoothly with
tinfoil, supported on a glass
stem. The end nearest
the glass cylinder carries a
rod terminated by a cross-
piece which is provided
with a number of fine brass
spikes, K. This is some-
times called the comb or
rake.

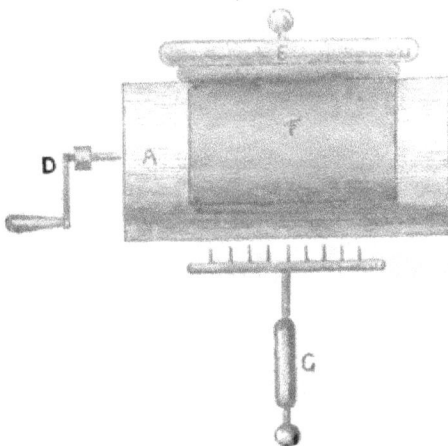

FIG. 109.

In fig. 109 we have the machine when viewed from above.

[1] Experience has taught us that a greater quantity of electricity is
developed by using a rubber coated with *electric amalgam.* An amalgam
is a combination of mercury and other metals. For electrical purposes it
generally consists of 1 part by weight of tin, 2 of zinc, and 6 or 8 of
mercury. Place the zinc and tin in a crucible, just melt, then add the
mercury. Stir while cooling. Reduce the mass to powder in an iron
mortar. It can be mixed with a little lard and applied to the rubber, or
the rubber may first be smeared with lard, over which the amalgam can be
sprinkled.

A cheap and useful cylinder machine may be made as follows :—

Take a good-sized glass bottle ; fit a metal rod, bent twice at right angles (so as to serve for a handle) into the stopper ; fit a straight rod into a piece of wood cut to fill the cavity at the bottom of the bottle.   Cement [1] both the stopper and the wood into their places.   Fix two uprights firmly into a wooden base, and upon these support the bottle by means of the rods. Cover a wooden cylinder having rounded ends, with tinfoil, and along one side of it place a row of pins, having previously filed off the heads.   This forms the prime conductor, which is then supported on a varnished glass stem.   A leather cushion may be stuffed with horsehair or wool, and a strip of silk sewn on.   This cushion, forming the rubber, is fastened to a wooden stem in such a manner that it presses upon the glass.

Action of the Machine.—If the rubber be insulated it must be first put into earth-communication by means of a metallic chain.   When the machine is in motion, the friction of the amalgam-covered rubber upon the glass generates electricity, positive being produced on the glass, and negative on the rubber.

(*a*) The prime conductor becomes positively charged in the following manner.

The positive electrification on the glass is carried round until it comes near the prime conductor.   This then causes negative electrification

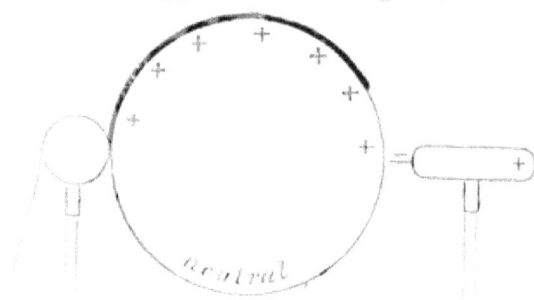

Fig. 110.

---

[1] The cement may be prepared by melting together 2½ oz. of resin, ½ oz. bees' wax, ½ oz. Chinese red, 1 dram plaster of Paris.   If, however, really good bees' wax be used, half the quantity would be better.   The addition of a small quantity of gutta-percha renders the cement tougher and more enduring.   The resin should be melted first in an iron vessel, and the wax added afterwards ; the plaster and colouring matter may then be stirred in.   The cement may be cast in a greased tube of paper.

to be induced on the near end of the prime conductor, and positive to be repelled to the further end. The negative electrification is discharged from the points, electrifying the air between them and the cylinder, which is, therefore, repelled upon

the glass, neutralising its positive electrification. Thus the glass arrives at the rubber in a neutral state, when it is again ready to be similarly acted upon. We therefore learn that the conductor is electrified posi-

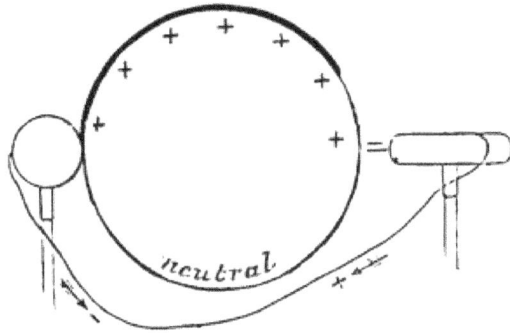

FIG. 111.

tively by losing its negative charge.

The action will be understood by aid of fig. 110, which represents a vertical section of a cylinder machine.

(*b*) If a negative charge be required, the rubber is insulated and provided with a metal knob at the back. The prime conductor must be placed in earth-communication.

(*c*) If both the rubber and the prime conductor be insulated, and then joined by a wire, no free electricity is manifest, for the positive and negative charges are equal in quantity, and therefore neutralise each other (fig. 111).

The Plate Electrical Machine differs from the

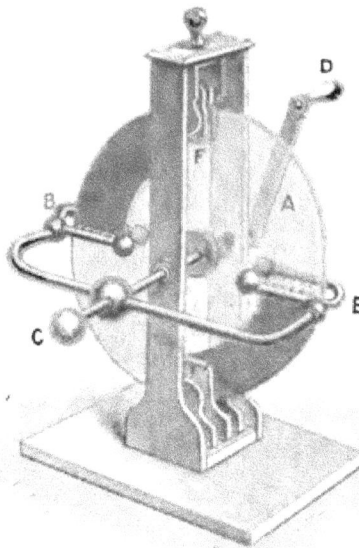

FIG. 112.

machine just described by having (1) a circular *plate* of glass or ebonite, A (fig. 112), instead of a cylinder. Ebonite has the advantage of being easily electrified, less liable to break and less hygroscopic.   Its disadvantage is, that its surface slowly changes so as to become a conductor.[1]   The plate revolves between two . pairs of *rubbers*, each, therefore, consisting of double cushions, one pair, F, fixed at the top, and the other F' at the bottom of the plate.   The *prime conductor* consists of a curved rod, to the front of which a knob, c, is attached.   The other end is termi-nated in the horseshoe-shaped rod, furnished with rows of spikes, between which the plate revolves.

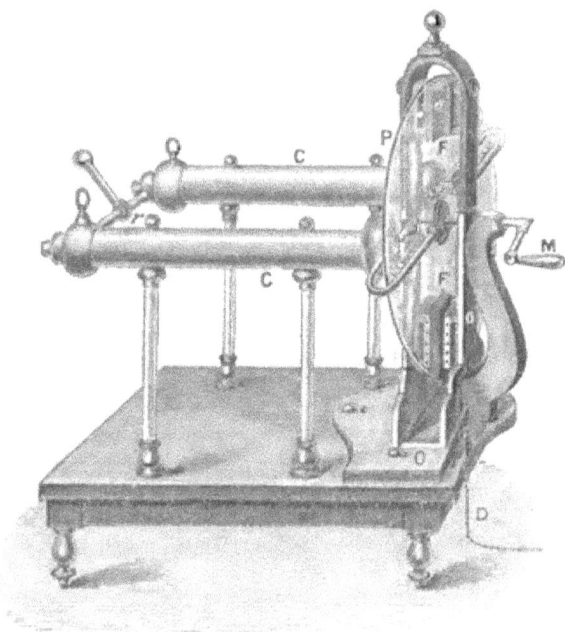

FIG. 113.

Another form of the plate machine is shown in fig. 113. In this machine the *prime conductor* consists of two parallel metal cylinders, c c, connected by a small rod, *r*. The other parts are similar to those in the one just described.

The *action* of the plate machine is precisely the same as that of the cylinder machine.

Winter's Plate Machine.—This machine (fig. 114) is of

---

[1] To remedy this defect, see foot-note on p. 86.

more modern form, and differs from the ordinary plate machine
in having (1) only one pair of
rubbers, furnished with a silk
flap ; (2) a spherical conductor
c, to which are attached two
wooden rings, one on each side
of the plate. The part of the
rings turned towards the plate
contains a groove lined with
metal from which a number
of fine spikes protrude. (3) A
large brass hoop enclosed in
well-baked polished wood (2
or 2½ feet in diameter), is often
fixed into an aperture in the
prime conductor. This ring
increases the surface, and
therefore the capacity, of the
conductor. Very long sparks,
three or four times the length

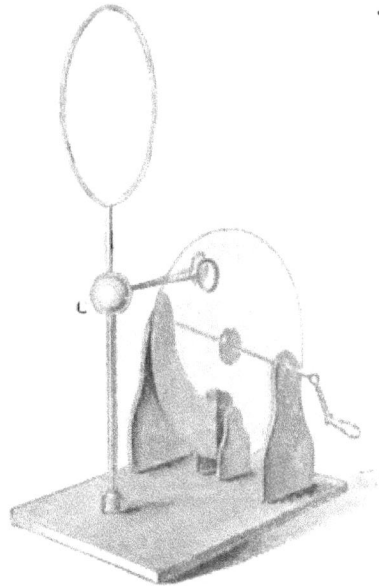

FIG. 114.

of those from an ordinary machine, can be obtained.

Several machines have recently been constructed, e.g. those

FIG. 115.

of Holtz, Voss, and Wimshurst, which are to a great extent

I

superseding the older forms. They are **exceedingly convenient** for obtaining a supply of electricity, but the theory of their action is rather too difficult for this work. For this reason, merely a short description will be given of the last-named machine.

Wimshurst's **Machine** consists of two varnished glass or ebonite plates, revolving in opposite directions, upon which strips of tinfoil are gummed. In fig. 115 the unshaded strips are on the front of the nearer plate, and the shaded ones on the back of the farther plate. Two conductors lie obliquely across the plates, at right angles to each other, one at the front and the other at the back. These conductors are terminated in brushes, which, as the plates revolve, come in contact with the strips. The discharging part of the machine is in connection with two insulated horizontal combs. The distance between the knobs can be regulated by means of the insulating handles shown on the right and left of the diagram.

Precautions to be observed in using the older forms of electrical machines.

(1) Glass being hygroscopic, the *moisture* which accumulates on the surface must be got rid of by thoroughly warming both the glass and the rubber. This is best done by placing the machine near, but not too near, a fire, and rubbing it with a piece of hot silk. The rubbers may be removed, if necessary, and dried before the fire. Too much stress cannot be placed on this point. The glass stems should be coated with shellac varnish.

(2) They must be free from *dust*.

(3) *Amalgam* should be occasionally spread over the rubbers.

(4) All unnecessary *points* must be carefully avoided.

Experiments with the electrical machine.

Exp. 135. If the rubber of a machine be insulated, connect it with the earth. On turning the handle the prime conductor will become charged (as described on p. 110). Present the knuckle to the prime conductor. Notice that the spark is sharp and stinging.

The reason that a spark is produced is as follows :—

The positive charge on the prime conductor acts by induction on the hand, attracting a negative charge and repelling a positive one. When the attraction between the positive charge on the conductor and the negative charge on the hand is sufficient to overcome the resistance of the intervening air, they neutralise each other, producing the spark.

Exp. 136. Hold a metal rod terminated in a knob at various distances from the prime conductor (fig. 116).

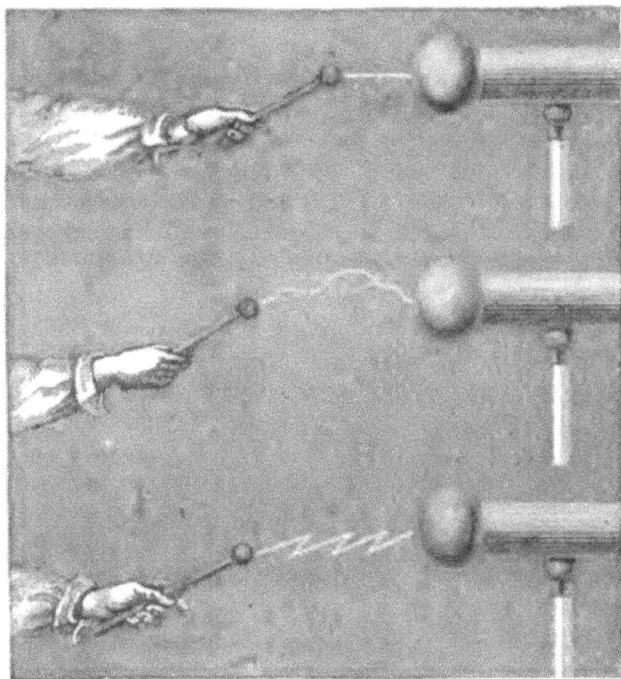

FIG. 116.

(*a*) at a short distance the spark is straight.

(*b*) at a distance of two inches or so, it becomes irregular like the branch of a tree.

(*c*) If the discharge is a powerful one the spark is zigzag.

These discharges are excellently seen with a Voss or Wimshurst machine.

Other luminous discharges will be given in Experiments 175, 176.

Exp. 137. Insert a Henley's quadrant electrometer into the prime conductor. Work the machine : notice the rise of the pith ball and its position on the graduated scale. Cease turning ; the pith ball gradually falls.

The reason is that the charge is continually escaping from the surface of the apparatus, owing to various causes, e.g. moisture and dust.

Exp. 138. Again work the machine ; the pith ball rises. Present a knuckle to the prime conductor : notice the rapid fall of the pith ball, owing to the neutralisation of the conductor.

Exp. 139. Make a tassel of fine shreds of tissue-paper. Attach it to the prime conductor by means of a bent wire. On working the machine the strips repel one another.

Heads of hair are sold to illustrate the fact that repulsion takes place between bodies similarly charged. They are placed on the prime conductor by means of a brass rod fixed into the head.

Exp. 140. Bore a hole in the bottom of a metal vessel—a tin canister answers well. Hang it to the prime conductor of a machine by means of a piece of moistened string tied round the top. Fill the vessel with water. Before the machine is turned the water

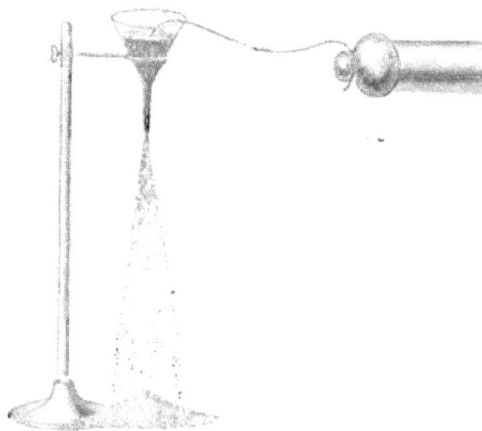

Fig. 117.

falls drop by drop. On electrification, the drops succeed each other with great rapidity.

Exp. 141. Fill a small glass funnel, having an aperture in the stem of $\frac{1}{8}$ inch with fine sand. Before electrification, it runs out in a fine stream. Pass a wire in connection with the prime conductor into the sand. On working the machine, the particles are scattered as they fall (fig. 117).

The reason of the results obtained in Exps. 139–141 is obvious.

The '*electrical chimes*' is an apparatus which admirably shows the phenomena of attraction and repulsion. It consists of three brass bells (fig. 118), suspended from a brass cross-piece, the two outer ones being suspended by a metallic chain, the central one by a silk thread. The latter bell is placed in earth communication by means of a chain. Two small brass balls hang by silk threads between the bells.

Exp. 142. Attach the apparatus to the prime conductor of a machine. On working the machine, the two outer bells, being in metallic connection with the conductor, receive a positive charge. The small balls are attracted, and by contact receive a positive charge; they are therefore repelled (and are also attracted by the central bell, which has been negatively electrified by the inductive action of the outer bells). On contact with the central bell the balls

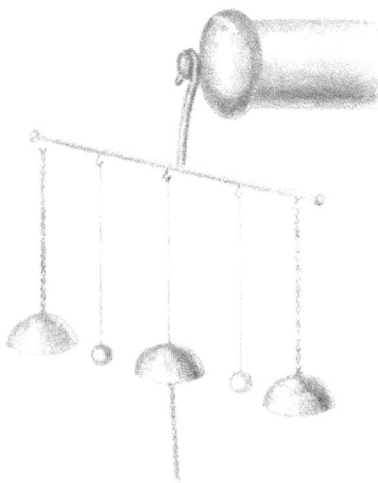

Fig. 118.

lose their charge, and are therefore again attracted by the outer bells. These actions are repeated while the machine is in motion.

Exp. 143. Place some pith balls upon an *uninsulated* metal plate. Another plate is supported on a glass stem, and is brought above the first one (fig. 119). Connect the upper plate with the prime conductor of a machine. On turning the machine, the pith balls are (1) attracted, (2) electrified by contact, and (3) repelled. They lose their charge on coming in contact with the lower plate, and are therefore in a condition to be again attracted, and so on.

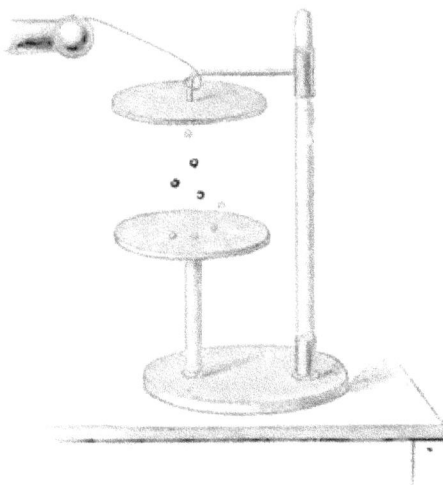

Fig. 119.

Exp. 144. Let a person stand on an insulating stool, and place his hand on the prime conductor of a machine. Set the machine in motion. As the human body is a conductor, the following effects may be produced :—

(1) If the hair be dry, it will stand erect.

(2) Sparks can be drawn from any part of the body.

(3) A gas jet may be lit, if either the knuckle or a brass rod held in the hand be presented to the escaping gas.

(4) A balanced lath, or any light body, can be attracted if the hand be held near it.

EXERCISE XXII.

1. If the stand upon which the pith-balls are placed in Exp. 143 be made of vulcanite instead of brass, and the upper plate be electrified, the balls are attracted and then repelled, but they are not attracted again. Explain this.

2. How would you charge two insulated metal spheres, one positively and the other negatively, from an ordinary electrical machine?

3. Could an electrical machine be made to act if it had a brass plate instead of a glass one. If so, how? If not, why not?

4. When the handle of an ordinary frictional machine is turned, sparks can be drawn from the prime conductor. Carefully explain how the prime conductor becomes charged with electricity.

Further experiments with the electrical machine, showing the action of points.

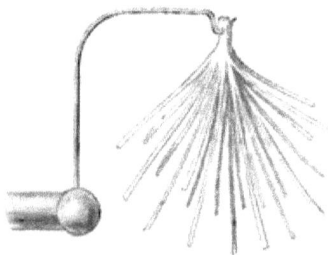
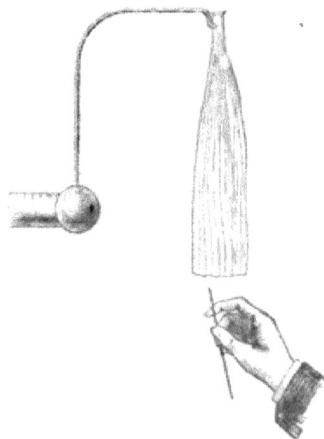

FIG. 120.   FIG. 121.

Exp. 145. Fix a Henley's quadrant electrometer on the prime conductor. Hold a needle towards the conductor, and set the machine in motion. Notice that the pith ball remains at rest.

The electrified conductor acts by induction on the body holding the needle, attracting a negative charge, which is discharged from the point, neutralising the positive electrification of the conductor.

Exp. 146. Repeat Exp. 139. Hold a needle, the point of which is covered with the finger, under the strips (fig. 120): they stretch towards the hand. Uncover the needle-point: observe that the strips collapse (fig. 121). The reason is that the induced negative charge passes from the point, neutralising their positive charge.

Exp. 147. Fit an 'electric whirl' (fig. 122) into the prime conductor of a machine. This little apparatus (sometimes also called *Hamilton's mill*), consists of pointed wires bent at right angles, and arranged in such a manner that the points are all in the same direction. The wires fit into a brass cap, which is supported on a pointed brass rod. Turn

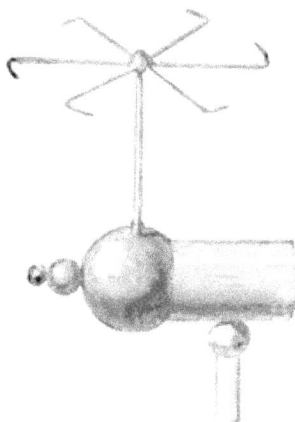

FIG. 122.

the handle of the machine. Notice that the whirl rotates in a direction opposite to that in which the points are turned.

The positive electricity is discharged from the points; the surrounding particles of air are therefore electrified with a similar charge; repulsion takes place between them and the whirl, which is in consequence set in motion. The currents of air can readily be felt by the hand.

Electrical Inclined Plane.

—An interesting experiment may be performed by means of the following cheap and simple apparatus. Bend

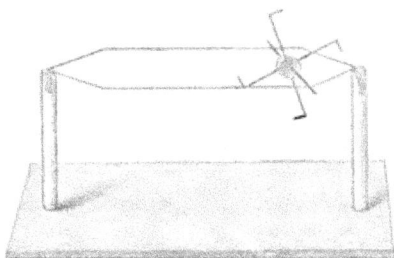

FIG. 123.

two copper wires as shown in fig. 123. Insert the ends into two varnished glass tubes, one of which is a little shorter than the other. Pass a needle through a cork. The needle forms an axle which rests on the two copper wires. Take four small pieces of wire; bend each wire so as to form a right angle, one arm of which is longer than the other. File the short arm to a point, and insert the other end into the cork, arranging them as shown in the diagram. Take care that the points are all turned in the same direction.

Exp. 148. Fasten a wire from the prime conductor of an electrical machine to one of the copper wires. On working the machine the whirl rotates, and at the same time moves up the inclined plane. The reason of this will be easily understood.

Exp. 149. Fix a pointed wire to the prime conductor as shown in fig. 124. Hold a lighted candle near the point. Notice that

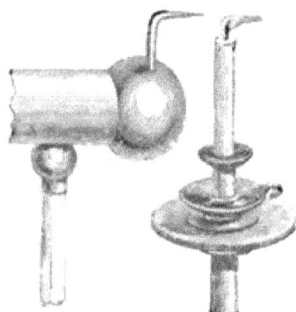 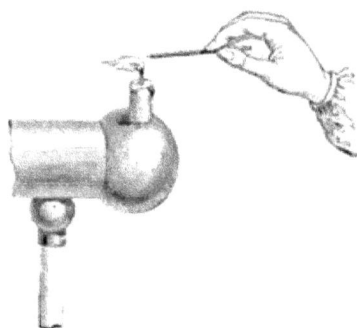

FIG. 124.   FIG. 125.

on working the machine the flame of the candle is blown from the point, owing to the outward rush of the repelled air.

Exp. 150. A similar effect is produced if a candle be placed on the prime conductor (fig. 125) and a pointed wire be held in the hand. The flame is again blown away from the point.

Exp. 151. Place a gold leaf upon a sheet of paper. Electrify a vulcanite rod. Allow the gold leaf to drop from the paper. Bring the electrified rod near.

Notice (1) violent attraction of the leaf.

          (2) sudden stoppage before contact.

          (3) repulsion.

Follow the leaf with the rod. It may be retained in the air for a long time.

The action is as follows :—A positive charge is induced on the gold leaf near the rod, and a negative charge repelled. The leaf is therefore first attracted, but the positive charge escapes from the edges and points near the rod, leaving the leaf negatively electrified. Therefore repulsion ensues.

1. An orange, into which a sewing needle has been stuck point outwards, is suspended by a dry silk thread. A charged body is brought near to it, 1st, opposite the point of the needle ; 2nd, opposite the side remote from the needle. State and explain the electrical effect in each case.

2. An insulated and electrified conductor can be discharged by bringing near it the point of a sharp needle held in the hand. Explain this.

3. Explain the action of the *electric whirl.*

4. The prime conductor of an electrical machine has a long brass rod projecting from it. From the end of the rod a pith-ball hangs by a damp cotton thread, and a pin is driven into the pith ball up to its head, so that the point projects on the other side. How and why does the ball move when the machine is turned ?

5. A piece of bees' wax is stuck on the prime conductor of a machine. The eye of a needle is then passed into it, so that the needle touches the prime conductor. On working the machine scarcely any sparks can be obtained when the knuckle is presented to the conductor. Why is this ?

6. I insulate two metal balls, and place them two inches apart. I then fasten a needle (by means of bees' wax placed round the eye) on the far side of one of them. I bring a vulcanite rod, rubbed vigorously with flannel, between them. Draw and describe the result.

7. A sharp point attached to a conductor, A, is held near an insulated charged conductor, B. What will be the effect on B if A is (1) insulated, (2) uninsulated ?

Coulomb's Torsion Balance will now be described, by means of which two extremely important laws may be proved. The student is not, however, required to perform the experiments, as great skill is requisite to obtain anything like accurate results.

The first law is :—

The force of attraction or repulsion between two electrified bodies varies inversely as the square of the distance between them, i.e.

$$F \propto \frac{1}{d^2}$$

Where F = the force of attraction or repulsion, and d = distance.

The instrument consists of a light rod of shellac, p (fig. 126), having at one end a small metal disc, n, suspended within a cylindrical glass vessel, A, by means of a very fine silver wire. The upper end of the wire is fastened to a brass button, *t*, in the centre of a graduated disc, *e*, divided into 360°: *a* is a fixed index, to show the number of degrees the disc is turned. A gilt pith ball, *m*, is fixed on a glass rod, *i*, which passes through the aperture *r*. A scale, *oc*, divided into degrees, is fixed round the glass cylinder on a level with the pith ball.

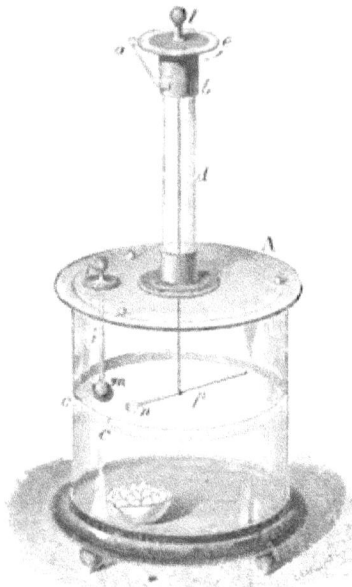

Fig. 126.

Method of using the torsion balance.—The torsion head is turned until the disc *n* and the pith ball *m* are in contact. The glass rod *i* is removed and the ball *m* is charged with electricity: the glass rod is then quickly replaced. When *m* and *n* touch, *n* receives part of the charge on *m*, and repulsion ensues. The wire becomes twisted : the *force of the twist*, i.e. the *force of torsion*, becomes greater the more the wire is twisted.

When *n* remains at a constant distance from *m*, read off the number of degrees on the scale *c*. Now *the force of torsion is proportional to the angle of torsion.* Suppose the angle is 20°, then the force of torsion is 20 times as great as it would have been if the angle had been 1°.

In a particular experiment Coulomb found the angle be-tween *m* and *n* to be 36° ; the twist on the wire was 36°, i.e., the force of the repulsion was 36 times as great as the force which would twist the wire 1°.

The torsion head was then turned so that the distance between *m* and *n* was halved, i.e. they were 18° apart. To do this it was found necessary to turn the disc through 126° in an oppo-site direction. The bottom part of the wire also had 18° of twist. The total twist was therefore 126° + 18° = 144°. To bring *m* and n 9° apart—i.e., to make the distance $\frac{1}{4}$—the total twist would have been 576° ; therefore

the distance being $1 : \frac{1}{2} : \frac{1}{4}$

the force of repulsion is $36 : 144 : 576$;

or, getting the simplest ratio by dividing by 36 we have

the force of repulsion $1 : 4 : 16$

which latter numbers are obtained by inverting the distance and squaring ; thus proving that the force of repulsion varies inversely as the square of the distance.

By a somewhat modified experiment the force of attraction between two oppositely electrified bodies may be proved to vary inversely as the square of **the** distance between them.

The second law was also **proved** by Coulomb by means of the same apparatus, viz :—

The **distance** remaining the same **the force of** attraction **or** repulsion **between** two small electrified bodies is proportional to the product of the quantity with **which** they are **charged.**

The two laws are included **in the equation**

$$F = \frac{q \times q'}{d^2}$$

where F = force ;

q = quantity of electricity in one charge ;

q' = quantity of electricity in the other ;

d = distance.

1. A small insulated **brass** ball has 20 units of positive electricity, and is placed 6 centimetres from a similar ball with 9 units. Find the force of repulsion.

$$F = \frac{20 \times 9}{6^2} = \frac{20 \times 9}{36} = 5 \text{ dynes.}$$

2. **Two balls are charged** with **+ 12** and **+ 8** units of electricity. With what **force** will they repel each other when **placed at** a distance of **4** centimetres from each other ? *Ans.* : 6 dynes.

3. The two balls are charged with + 20 and − 9 units of electricity. With what force will they attract **each** other when placed at a distance of 3 centimetres from each **other?** *Ans.* : 20 dynes.

4. Two small insulated brass balls are charged, one with 5 units of positive electricity. They repel each other with a force of 4 dynes when placed 5 centimetres apart. Find the charge on the other. *Ans.* : + 20 units.

Several sensitive and delicate electrometers—e.g., Sir William Thomson's quadrant electrometer, and the attracted disc electrometer—are now very largely used. For information respecting these instruments the student must read a more advanced work on the subject.

CHAPTER XIII.

CONDENSATION AND CONDENSERS.

ON each side of a plate of glass fasten sheets of tinfoil opposite
each other, so that a space is left between the edge of the tinfoil
and that of the glass. If the glass plate be approximately 12 inches
long and 10 inches wide, a two-inch margin may conveniently be
left. It is advisable to coat this margin with shellac varnish. The
glass is often placed in a wooden frame, one sheet of tinfoil being
connected with the frame by a strip of tinfoil.

Such an apparatus is called a **fulminating pane** or **Franklin's
plate.**

FIG. 127.

Exp. 152. Place one side in earth-connection by means of a
chain, or by merely placing the finger upon it. Electrify the other
side as shown in fig. 127. Remove the earth-connection and set
it on its edge. Touch the sheet of tinfoil, which has been in
connection with the machine, then the other sheet, and so on alter-

nately. A very slight shock is felt each time the tinfoil is touched. An explanation of this will be given in Exp. 159.

Exp. 153. Now place the pane flat upon a table, and recharge. Put a finger of one hand on the lower sheet, and touch the upper one with the other hand. A violent shock is felt.

Exp. 154. The last experiment may be varied in an amusing way by placing a coin on the upper surface. After charging, request someone to pick the coin up ; the shock is so violent that the removal is rarely effected on the first trial.

Let us examine the reason of these results :

Fig. 128 is a sketch of a section of a Franklin's plate. The tinfoil A is charged positively from the prime conductor of a machine ; induction takes place across the glass, negative electricity being attracted, and positive repelled to the earth. This negative on B reacts on the positive on A, attracting and accumulating it on the side next the glass ; thus the side next the prime conductor is rendered capable of receiving a further charge. This in its turn attracts more negative electricity on B, and so on ; we, therefore, learn that the accumulation of the negative charge on B continually increases the capacity of A until a certain point is reached (see p. 106).

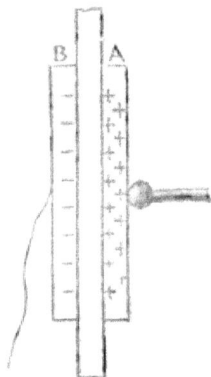
FIG. 128.

We thus see that the sheets of tinfoil can receive a greater charge by arranging them in this manner, i.e. the capacity of an insulated conductor is increased if another conductor be brought near, especially if the latter is earth-connected.

In Exp. 152 the opposite charges next the glass have thus reacted on each other, and are, therefore, ' bound.'

In Experiments 153, 154, when both sides are touched, the discharge passes through the body and causes a violent shock.

Exp. 155, to show the ' binding ' effect on the action of two charges.

Place a large sheet of tin (fig. 129) on an earth-connected copper wire lying on the table. Upon this place a smaller sheet of varnished glass, provided with two silk loops, and upon this a still

smaller sheet of tin, which is connected by means of a wire with
the disc of a gold-leaf electroscope. Electrify a proof-plane by

placing it on the prime conductor
of a machine at work. Touch the
upper sheet of tin with the proof-
plane. Repeat this several times,
and notice that there is no divergence of the gold leaves of the
electroscope. The charge is thus
bound by the attraction of the
induced opposite charge on the

FIG. 129.

lower sheet (as already explained). Now lift the glass by the
loops, thus removing the upper sheet from the inductive influence
of the lower, and the gold leaves at once diverge.

The reason of this will be clear from the following consideration. We have learnt that $c = \dfrac{Q}{v}$ (see p. 106)

$$\therefore v = \frac{Q}{c}$$

By removing the upper sheet from the influence of the
lower, the capacity is made very much less, and as the quantity
remains the same, the fraction $\dfrac{Q}{c}$ —i.e., the potential— is very
greatly increased. There is, therefore, a transference of electricity from the tin to the electroscope, which causes the
divergence of the leaves.

By thus employing two conductors separated by a non-
conductor, the charge is *condensed* or *accumulated*, and the
apparatus is called a *condenser* or *accumulator*.

Condensers may vary in form, but they always consist of

(1) two conductors separated by

(2) a dielectric (non-conductor).

The principle and action of a condenser will, however, be
more easily and fully understood by considering the following
apparatus.

Epinus's Condenser consists of two brass discs, A and B
(fig. 130), on insulating glass legs with a plate of glass, C, between
them. The brass plates are capable of being moved backwards
and forwards, and each is provided with a pith-ball pendulum,

which may be situated as in fig. 130, or which may be attached
to the back of the plates by means of bees' wax and a fibre of
silk.

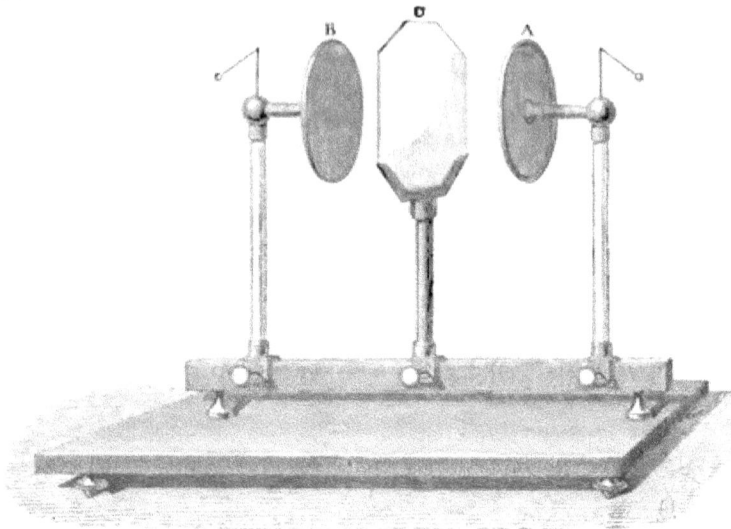

FIG. 130.

Exp. 156. By means of wires connect A with the earth and B
with the prime conductor of a machine (fig. 131). Work the

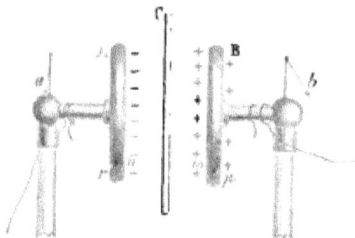

FIG. 131.

machine. Remove A to such a
distance that B has practically no
inductive action upon it. Notice
that the pith ball on B rises.

· Gradually bring A towards B.
Notice the gradual fall of the pith
ball. The positive electricity on
B induces a negative charge to
the face *n*, and repels a positive
charge to the earth. The negative charge on A, in its turn, acts
on B, disturbing the charge, and attracting more positive to the
side *m*, thus 'binding' it. Thus there is less positive electricity
on the side *p*, and the pith ball falls. B, having its capacity increased, can therefore receive an extra charge from the machine.
This extra positive charge reacts on A, more negative being attracted to *n*, which again reacts on B, still further increasing its
capacity, and so on.

Exp. 157. An excellent illustration of electricity being 'bound'

by inductive action may be given by means of a condenser and two
gold-leaf electroscopes (fig. 132). Move the plates of a condenser
some distance apart. Connect them by means of wires to the discs
of two gold-leaf electroscopes. Charge A positively and B nega-
tively (it is best to do this by induction).

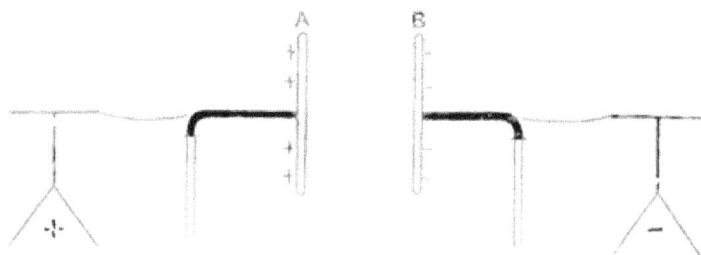

FIG. 132.

The charges spread over the conductors, and the **gold leaves**
will diverge as shown in fig. 132. Gradually bring the plates nearer
together, and notice the gradual collapse of the leaves. As A and
B approach, they come under each other's influence; inductive
action is set up, and the two dissimilar charges **attract each** other.
As the distance between **the plates diminishes, the action becomes**
gradually greater, the **charges accumulating on the plates, and**
therefore the gold **leaves collapse.**

As previously mentioned, this **experiment** is more accurately
explained by saying that, on **the approach** of the two plates,
induction takes place, by means **of which** the capacities of the
plates are increased, and the potential **is** therefore diminished.
This diminution becomes apparent by the collapse of the leaves.

Exp. 158.—Slowly separate the plates; **notice** the gradual
divergence of the leaves. **The** inductive action becomes less, **the**
capacities are diminished, **and** the potential **therefore** increases.

Exp. 159. Place the discs (fig. 131) in contact with the glass.
Connect B with the **machine and keep** A insulated. Electrify, **and**
observe the **rise** of both pith balls.

(1) Touch A with the finger. **Notice** the spark, and fall of **the**
pith ball *a*, **due to** the removal **of the free** positive charge.

(2) Disconnect the wire passing from the machine to the plate
B by means of a non-conductor. The pith balls remain as in (1).
B has a positive charge, of which part is accumulated on the side
next the glass, and part 'free' on the other surface.

K

(3) Touch B with the finger ; the free positive charge (i.e., the electricity which B would contain without the aid of the induced negative charge on A) escapes, and the pith ball *b* falls. Notice the rise of the pith ball *a*, due to the fact that some of the electricity becomes free, i.e. the potential of A rises.

(4) Touch A. Notice the fall of the pith ball, owing to removal of its free electricity, and the rise of that on B.

(3) and (4) may be repeated many times if the apparatus is perfectly dry.

The student should notice that this experiment is similar to Exp. 152.

This method of discharging the condenser is called the *slow discharge*.

FIG. 133.

The **discharging tongs**, or discharger, consists of two curved brass rods terminated in knobs, and joined by a hinge attached to one or two insulating handles. Fig. 133 represents the discharging tongs fitted with two glass handles. It is used as indicated in Exp. 160, to discharge a condenser with safety.

A cheap and useful form of discharging tongs may be made by bending a piece of stout wire as shown in fig. 134. Insert the ends of the wires into small wooden balls coated with tinfoil. Fasten the wire into a test-tube with hot shellac.

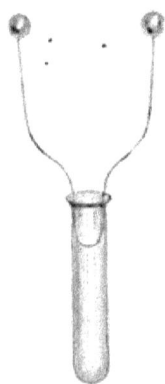

FIG. 134.

Exp. 160. Charge the condenser as in Exp. 156. Place one knob of the discharging tongs on one plate, and bring the other knob near the other plate. Notice that when the knob is some distance from the plate a sharp crack is heard, and a spark passes, due to the neutralisation of the two opposite charges. This method of discharging is called *instantaneous discharge*.

Thus there are two methods of discharging a condenser :—

(*a*) Slow discharge;

(*b*) Instantaneous discharge.

The latter takes place whenever a conductor connects the two plates of a condenser, e.g. in Exp. 153, where the charges neutralised each other through the body, and in Exp. 160, where they neutralise through the tongs.

1. Describe Franklin's plate, and explain its action.
2. Describe an experiment which will illustrate the action of the electric condenser.
3. I place a sheet of vulcanised india-rubber on a sheet of tinfoil, and a second sheet of tinfoil upon the rubber. I connect the lower sheet with the earth, and the upper one with the conductor of an electrical machine. What occurs when the machine is set in motion?
4. Why do the two plates of an air-condenser, if oppositely electrified, show less signs of electrification when placed near each other than when they are placed further apart?
5. Describe and explain the two methods of discharging a condenser.

Discovery of the Leyden Jar.—Exp. 161. Take a bottle similar to the one shown in fig. 135. Partly fill it with water. Into the neck of the bottle insert a cork, through which passes a nail of sufficient length to pass below the surface of the water; charge the bottle by holding the outside in the hand, and the head of the nail to the prime conductor of a machine in motion. Still clasping the outside, touch the nail with the knuckle of the other hand. A shock is felt.

You have made, and performed an experiment with, the first form of the Leyden jar, discovered by Kleist, a bishop of Cammin in Pomerania, A.D. 1745. In the following year Cuneus of Leyden made a somewhat similar dis-

FIG. 135.

covery. Its present construction is due to Dr. Watson, Bishop of Llandaff, who coated the interior and exterior of the bottle with silver-foil.

To **make a Leyden jar.**—Obtain a wide-mouthed glass bottle about 6 or 7 inches high—a pickle-bottle answers very well. First paste tinfoil on the inside of the bottle, leaving $1\frac{1}{2}$ or 2 inches of glass uncovered at the top. It is advisable to cut the tinfoil into strips and place them one by one down the

sides, allowing them to overlap slightly; a circular piece is then
placed over the bottom. Another excellent plan for coating
the interior is to cut a strip of tinfoil of the same size as the
exterior coating, gum or paste the bottle and roll up the tinfoil,
and drop it into the bottle; unroll it, taking great pains to secure
a smooth surface (a piece of wood, enlarged at one end, and
covered with linen, is useful for this purpose); as before, place
a similar piece at the bottom. Next cover the exterior with
tinfoil, leaving a similar margin at the top. The glass margin
should then be covered with shellac varnish. Fit a wooden

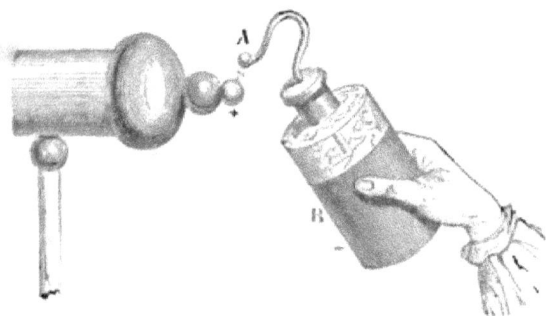

FIG. 136.

stopper into the
mouth, having first
passed a stout
brass wire through
the centre. A
brass knob or
leaden bullet must
be fastened to the
end of the wire
outside the jar,
and to the inner
end a chain is fastened, which must lie on the bottom of the jar.

A Leyden jar is said to be charged with the kind of electricity

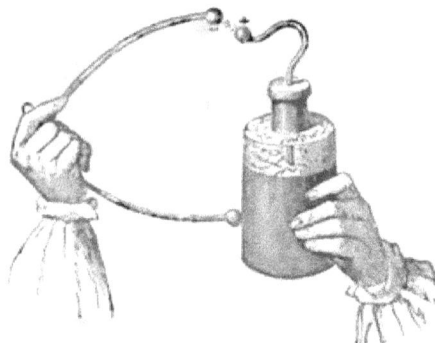

FIG. 137.

accumulated on the inner
coating.

Exp. 162, to charge a
Leyden jar positively. Hold
the outer coating in the hand,
and present the knob to the
prime conductor of an electrical machine in motion
(fig. 136). Discharge it instantaneously by (1) putting
the knob of the discharging-
tongs to the outer coating,
and (2) bringing the other knob near the knob of the jar (fig. 137).

Exp. 163. Recharge as in the previous experiment, and place
the jar on the table. Allow a number of persons to form a circuit by

joining hands. The person at one end holds the outer coating, and the one at the other end touches the knob. A shock will be felt through the circuit.

The action of the Leyden jar is precisely similar to that of the fulminating plane, or of Epinus's condenser.

Exp. 164, to prove that electricity passes from the outer coating. Stand on an insulating stool, hold the outer coating in the hand, and present the knob to the prime conductor of a machine in motion. Notice effect of bringing the knuckle of the other hand near (1) the electric pendulum, (2) the disc of a gold-leaf electroscope, (3) a balanced lath. Thus the person holding the jar is seen to be electrified.

Exp. 165, to prove that this electrification is positive. Charge a gold-leaf electroscope positively by induction. As in the last experiment stand on an insulating stool, and hold the outer coating of the Leyden jar in one hand, and bring the other near the disc of the electroscope. The leaves diverge more. The inducing body is therefore positively electrified.

Exp. 166. The same result may be obtained by insulating the Leyden jar, and connecting the outer coating by means of a wire with a gold-leaf electroscope. Turn the machine very carefully, and notice divergence. Remove the wire by a non-conducting rod.

Test the nature of the charge in the leaves by means of a positively electrified body.

Exp. 167, to charge the jar negatively. (*a*) Hold the knob of the jar in the hand, and bring the outer coating to the prime conductor of an electrical machine. On turning the machine a positive charge enters the outer coating, which acting by induction through the glass attracts negative electricity, and repels positive, which escapes to the earth.

(*b*) One knob of a Wimshurst machine gives positive electricity and the other negative. The knob of the jar may be held, in the ordinary way, to the negatively charged knob.

(*c*) With an ordinary plate or cylinder machine, if the prime conductor be earth-connected, and the rubber insulated, a negative charge may be obtained from the rubber.

Exp. 168 is a very interesting one, to illustrate the opposite electrical conditions of the two coatings of a Leyden jar. Charge the jar positively as in Exp. 162.

(1) Holding the outer coating, draw a series of lines with the knob on a cake of well-dried vulcanite.

(2) Place the jar on an insulator, and then take hold of the knob and trace another series with the outer coating.

(3) A mixture of red lead and flowers of sulphur is shaken through a muslin bag from a height above the cake. Owing to the friction caused by shaking the red lead and sulphur together, the red lead becomes positively, and the sulphur negatively charged. The red lead therefore seeks the lines traced by the exterior coating of this jar, and the sulphur the lines traced by the knob.

Such figures are known as ' Lichtenberg's figures.'

Fig. 138 represents the result of a particular experiment. The central cross was drawn on a circular piece of vulcanite with

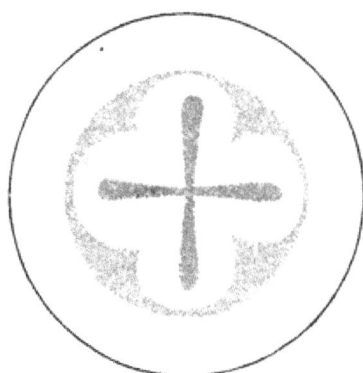

FIG. 138.

the outer coating of a positively charged Leyden jar; the red lead, therefore, accumulated there : the circle was drawn by the knob ; the sulphur fell as shown. The selection of the red lead for the negative cross, and of the sulphur for the positive circle, is due to both attraction and repulsion, i.e. the negatively charged sulphur was attracted by the positive circle, and repelled by the negative cross. Now, as there is less repulsion in the space between the arms of the cross than at the ends, the particles of sulphur arranged themselves as indicated in the figure.

Seat of Charge.—Exp. 169. Take a glass vessel B (fig. 139) having two movable metallic coatings, C and D. Place the parts together so as to form a complete Leyden jar. Charge it in the usual manner. Remove the inner coating with a non-conducting rod, or by the glass tube which is often placed round the stem. Then lift the glass vessel from the outer coating. As each part is removed bring it near the disc of an uncharged gold-leaf electroscope. The coatings produce no disturbance of the leaves ; the glass, however, at once causes the leaves to diverge. Put the jar together again. It may now be discharged by means of the discharging tongs.

Why is this? The fact is that the coatings act chiefly as conductors, and the charges reside on the inner and outer surfaces of the glass.

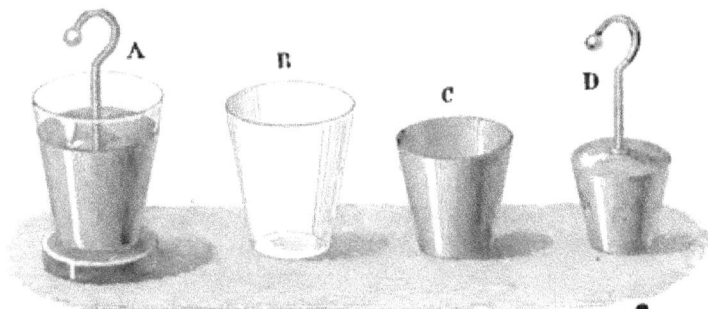

FIG. 139.

Capacity **of a condenser.**—The capacity of a condenser depends upon

(1) the size and form of the metallic conductor,

(2) the nearness of the plates, i.e. the distance across the dielectric,

(3) the inductive capacity of the dielectric.

As pointed out under 'inductive capacity,' p. 85, the medium—i.e. the dielectric through which induction takes place—performs an important part in the action. The power with which dielectrics allow induction to take place through them is called their *inductive capacity.*

Exp. 170. Suspend a brass ball by a silk ribbon. Electrify the ball, and bring it near the disc of a gold-leaf electroscope. Notice that the leaves diverge. Place a dry glass plate between the ball and the disc. There is an increased divergence. Glass has therefore a higher inductive capacity than air.

According to recent investigations the following bodies are placed in the order of their inductive capacity: glass, shellac, sulphur, ebonite, paraffin, air.

Limit of the charge of condensers.—There is, however, a limit beyond which a condenser cannot be charged. When dielectric polarisation (see p. 86) takes place, the medium is put into a state of strain (see footnote, p. 100), from which it continually endeavours to free itself. Thus in a charged Leyden jar the stress is produced by the opposite charges causing a strain in the glass ; in the air condenser (Epinus's condenser without the glass plate) the air between the two plates is strained. When the strain becomes too great—i.e. when it exceeds a certain limit—a discharge occurs across the dielectric ; e.g., if a Leyden jar be highly charged, and the glass be too thin, it is liable to be fractured by the attracting charges. This is called disruptive discharge.

Residual Charges.—Exp. 171. Charge and discharge a Leyden jar as in Exp. 160. After allowing it to stand for a short time, place one knob of the discharger to the outer coating, and the other to the knob of the jar. Observe that a second discharge takes place. If the jar be absolutely dry, a third, fourth, or even fifth discharge may be obtained. These are called *residual charges*, and are due to the fact that the charges have penetrated below—as it were, soaked into—the surface of the glass by their attractive action.

During the first discharge, the largest portions of the charges neutralise each other, but those which have penetrated into the glass require a little time to come to the surface. A more correct way of explaining them is, however, on the supposition of the electric strain ; the molecules of the glass, being acted upon by the stress of the opposite charges, are strained, and are therefore unable to recover their original form and volume immediately.

Leyden Battery.—As the capacity of a Leyden jar depends upon (α) the size of the coatings, it follows that a large jar can receive a much greater charge than a small one, and (β) upon the nearness of the coatings, it follows that, other things remaining the same, a greater charge can be given to a jar with thin glass, than to one with thick glass ; but, as already pointed out, thin glass is liable to be pierced by the discharge.

If powerful charges are required, a number of jars must be used, which may be arranged so that there is metallic connection between all the inner coatings, and also between all the outer coatings. This arrangement is shown in fig. 140.

The jars are placed on a conductor, e.g. a piece of tinfoil, which thus connects their outer coatings.

Their inner coatings are joined by means of a wire passing through holes in the knobs. The jars are commonly placed in

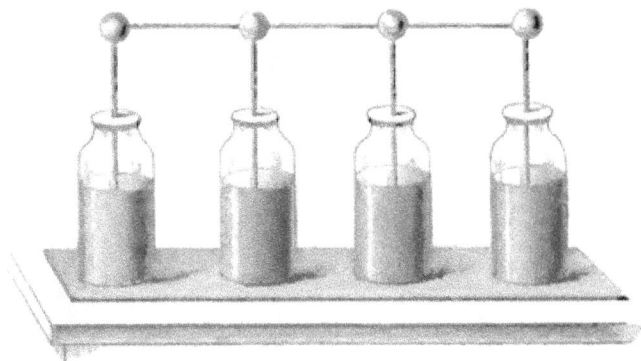

FIG. 140.

a box, the bottom being lined with tinfoil which is in connection with a hook or handle on the outside of the box. The inner coatings are connected as shown in fig. 141.

FIG. 141.

Universal Discharger.—This apparatus (fig. 142) consists of two movable brass arms, provided with universal joints, and supported on glass legs. The object to be experimented upon is placed on a small table made of hard baked wood, between two knobs, which terminate one end of each of the arms, the other end being terminated in rings.

FIG. 142.

Exp. 172. Place a lump of sugar, an egg, or a lemon on the table of a universal discharger. Connect the ring of one arm to the hook or handle on the outside of the box, by means of a wire—the discharger is now in metallic connection with the outer coating of the battery. Connect a knob of one jar with the prime conductor of a machine, and then charge the battery. Having darkened the room observe the luminous effects of the discharge on the sugar, egg, or lemon when the discharging tongs connect the other arm and the battery, as shown in the diagram. Great care must be taken in discharging a Leyden battery, as a serious, perhaps fatal, accident might occur.

Exp. 173. Place a small quantity of gunpowder on the table of the universal discharger ; fasten one end of a piece of *wet string* to the arm, and the other end to one knob of the discharging tongs. Discharge the battery, and the gunpowder will be fired. The wet string is necessary to retard the velocity of the discharge, otherwise the powder is merely scattered, as the spark passes too rapidly to ignite it.

The Cascade Arrangement.—In this arrangement, all the jars except one have their outer coatings insulated. The exterior of one jar is in metallic connection with the interior of the next (fig. 143). The outer coating of the last is in earth-connection. The interior of the first jar is connected with the prime conductor of a machine; on turning the handle the

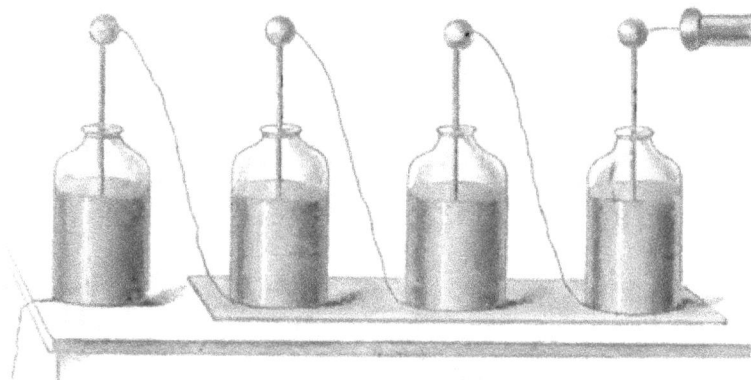

FIG. 143.

interior is therefore charged positively. Induction takes place, a negative charge being attracted, and the positive charge repelled, which escapes from the outer coating into the inner coating of the second jar; which, again acting by induction, repels a positive charge into the inner coating of the third jar, and so on through the entire series; the positive charge passing from the last jar to the earth.

Harris's unit jar.—This instrument is used for measuring the charge given to a Leyden jar. It consists of a small Leyden jar, A, about four inches long, and $\frac{3}{4}$ inch in diameter, fixed as shown in fig. 144 on an insulating stem, B. The rod P is connected with the prime conductor of a machine, and the outer coating with the jar or battery by the rod tp. When the electricity has accumulated to a certain extent on the inner surface of the jar, A, a spark passes between *m* and *n*. The distance between *m* and *n* may be varied by means of the ring

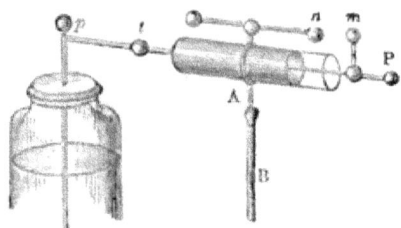

FIG. 144.

which slides along the jar. When the spark passes, a certain quantity of electricity is received as a charge by the battery, which we may call the unit. If, say, six sparks pass, we have six times the charge given by one spark ; thus the amount of electricity given to a jar can be calculated in terms of the small jar.

The Condensing Electroscope, invented by Volta, is an ordinary gold-leaf electroscope, provided with another disc, of the

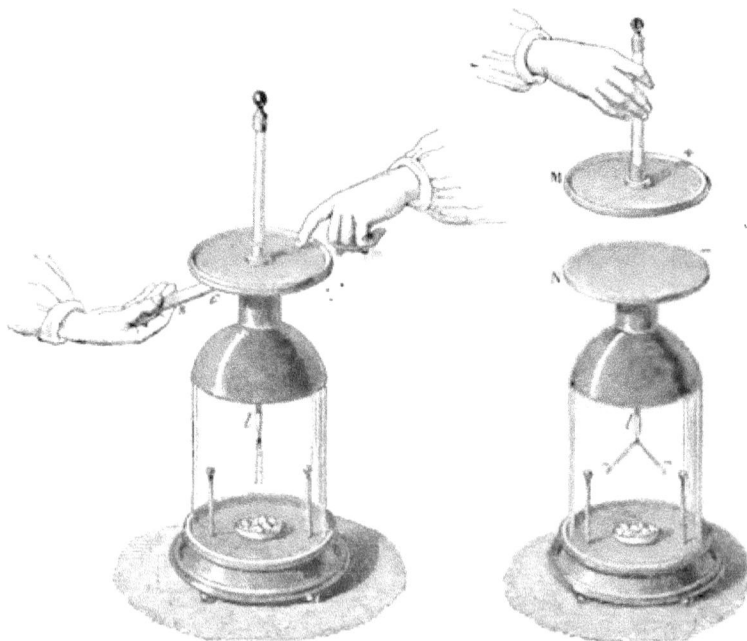

FIG. 145.

same diameter as that of the electroscope, to which is fixed an insulating handle. The faces of the two discs are coated with shellac varnish, which plays the part of the dielectric in a condenser. The main use of the condensing electroscope is to test electricity from a weak but continuous source.

The disc of the electroscope is called the collecting plate, and is touched by the body to be tested, while the other is called the condensing plate.

Exp. 174. Take a compound bar, made of zinc and copper soldered together. Hold the zinc end in the hand, touch the lower disc with the copper, the upper plate being touched by a finger of the other hand. Remove the hand and the rod, and lift the upper plate ; the capacity of the lower plate is greatly diminished, and the potential rises, which causes a divergence of the gold leaves. The charge will be found to be negative.

This experiment was devised by Volta to show that electricity was generated by contact of two dissimilar metals (see footnote p. 159).

That electricity can be developed by mere contact has, however, been proved once for all, by the following simple experiment of Sir William Thomson. He suspended a thin strip of metal so as to turn about the point A (fig. 146), and which is electrified with a known charge. Under it are placed two semi-circular discs or rings of dissimilar metals. No attraction takes place between either of the metals and the strip until the two dissimilar metals are placed in contact, when the strip at once turns, being attracted by

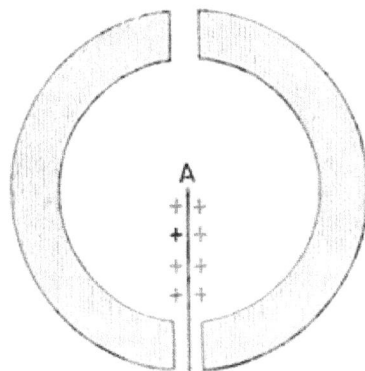

FIG. 146.

the oppositely electrified metal, and repelled by the similarly electrified one.

Contact Series.—If any two bodies in the following list be placed in contact, the first becomes positively, and the second negatively, electrified :—

+ Sodium
  Magnesium
  Zinc

Lead
Tin
Iron
Copper
Silver
Gold
Platinum
Graphite (Carbon)

EXERCISE XXVI.

1. Describe and explain the original experiment which led to the discovery of the Leyden jar.

2. Describe a Leyden jar, and the method of discharging it.

3. You are required to make a small Leyden jar. How will you proceed?

4. Explain fully what occurs during the charging and the discharging of a Leyden jar.

5. On touching the knob of a charged Leyden jar standing on the floor or on a common table, you get an electric shock ; but if either you or the jar stand on a dry cake of resin, you do not get a shock on touching the knob. Explain this.

6. What influences the amount of charge that can be given to a Leyden jar ?

7. How is electricity stored in a Leyden jar ?

8. Describe fully how you could charge a Leyden jar from the positive conductor of an electrical machine so as to get at will either a positive or negative charge on the inner coating.

9. One person holds a charged Leyden jar in his hand by its outer coating, and another holds similarly an uncharged jar. What happens when the knobs of the two jars are brought together ?

10. Two pith balls suspended, one by a damp cotton thread, the other by a dry silk thread, are each of them touched by the knob of a charged Leyden jar, which is held in the hand by its outer coating. Will there be any difference between the behaviour of the balls? If so, what difference, and why?

11. You have an electrical machine standing on a table with a glass top, and you have no means of connecting it electrically with the earth. What would you do in order to charge a Leyden jar by the machine?

12. A Leyden jar is made of gutta-percha instead of glass. Would any difference be produced in the effects?

13. Describe a Leyden jar with movable coatings, and say what it is intended to prove.

14. Describe a Leyden battery.

15. The inner coating of a Leyden jar is connected by a wire with the prime conductor of an electrical machine, and also with a gold-leaf electroscope. If the jar rests on a sheet of glass, a quarter of a turn of the machine produces a large divergence of the leaves of the electroscope. If the glass be removed, ten turns of the handle are required to produce the same divergence. Explain this.

√

CHAPTER XIV.

ELECTRIC DISCHARGE.

THE neutralisation of two opposite charges of electricity, which is known as **electric discharge**, is capable of producing a variety of effects. It is now proposed to study these in detail. The effects may be classed under the heads : (1) luminous, (2) heating, (3) chemical, (4) mechanical, (5) magnetic, (6) physiological.

(1) **Luminous effects.**—These experiments should be performed in a darkened room.

(*a*) Exp. 136 illustrated the *spark* discharge.

(*b*) **Exp. 175,** to illustrate the *brush* discharge. Attach a short wire—the free end of which has been rounded with a file—to the prime conductor of a machine. On working the machine a hissing sound is heard, quite different from the sharp crack in the spark discharge. This discharge is readily visible in the dark, and has a ramified, brush-like appearance (fig. 147). The brightness and size may be increased by holding a metal plate a short distance from the wire.

FIG. 147.

(*c*) **Exp. 176,** to illustrate the *glow* discharge. Instead of the wire having a round end, as in the last experiment, attach one with a sharp point. Notice the quiet and continuous pale blue glow.

The glow discharge is sometimes visible at the tops of the masts of ships when there is 'thunder in the air.' It is known amongst sailors as **St. Elmo's fire.**

In Exp. 172 we had another example of the luminous effects of discharge.

Exp. 177, to illustrate the influence of the pressure of air on the electric light. Obtain an apparatus known as the *electric egg* (fig. 148). This consists of an oval glass vessel, having metal caps at the ends, through each of which passes a rod terminated by a small ball in the interior. The lower rod is fixed, but the upper one moves up and down through a 'leather stuffing-box,' which renders it air-tight. The lower cap is provided with a stop-cock, and is so arranged that it can be screwed to an air-pump. Exhaust the air, and then screw the vessel into a heavy metallic foot, to which it is usually fixed. Connect the upper end with the prime conductor of a machine, and the lower with the earth. On working the machine the interior is filled with a feeble violet light. Gradually admit the air, and notice that the light becomes white and intermittent, having the ordinary spark-like form.

FIG. 148.

Exp. 178. Make a *luminous tube* (fig. 149) by taking a long glass tube (one or two feet in length), and gumming round it a number of lozenge-shaped pieces of tinfoil, arranged in spiral form, and having the points of each piece a short distance apart. The ends of the tube are terminated with brass caps, to which knobs may be attached. Hold one end in the

FIG. 149.

hand, and present the other to the prime conductor of an electrical machine in motion. Sparks pass between the points of the tinfoil, which give a brilliant appearance in a darkened room.

L

The beautiful effects obtained through Geissler's tubes will be treated of in Voltaic Electricity (p. 215).

(2) **Heating effects.**—Several experiments have already been shown which exhibit the heating effects of the discharge; e.g. Exp. 144, in which a jet of gas was lit, and Exp. 173, in which gunpowder was fired.

Exp. 179. Place a metal tray on varnished glass tumblers. Arrange two wires immediately over an unlighted gas-jet, with the ends $\frac{1}{10}$ of an inch apart. Connect one of the wires with the earth and the other with the tray. Rub hot brown paper vigorously with a clothes-brush and place it on the tray. The gas will be lit.

Exp. 180. Place a little ether in a shallow glass vessel—a watch-glass answers very well. Fasten a small brass knob to a chain about ten or twelve inches long. Place the knob in the bottom of the vessel containing the ether, and hold the other end of the chain in contact with the outer coating of a charged Leyden jar. Bring the knob of the Leyden jar near that in the vessel. A spark passes and the ether is fired. Fig. 150 represents the arrangement for this experiment.

FIG. 150.

(3) **Chemical Effects.**— **Exp. 181.** Soak a piece of blotting or filter paper with a solution of starch and potassium iodide. Place it on a glass plate, and holding one corner, connect the other with the prime conductor of a machine. On turning the machine a dark blue coloration, due to the liberation of iodine, will appear on the paper near the prime conductor.

If potassium iodide be used alone, the coloration will be brown.

Exp. 182. Attach two platinum wires to the conductors of a Voss or Wimshurst machine. Dip the free ends in a solution of copper sulphate. On working the machine, the wire in connection with the negative conductor will become coated with copper.

Exp. 183, to show the formation of water by the combination of its constituent gases—hydrogen and oxygen. Take a long, graduated, strong, glass tube, called a **eudiometer** (fig. 151), open at one end and closed at the other. Through the glass near the closed end are fused two platinum wires, the ends of which are near together in the interior. Fill the eudiometer with mercury, and invert it over a vessel containing the same liquid. Pass two volumes of hydrogen and one volume of oxygen into the tube, which should not be more than half full of the mixed gases, as great heat is evolved by their combination, which produces a great increase of volume. It is advisable to press the open end on an india-rubber pad placed at the bottom of the vessel. Connect *a* with the prime conductor of a machine, and *b* with the earth. On working the machine a spark passes, and a film of water can be seen on the inside of the tube. On raising the tube from the india-rubber there is a rapid rise of the mercury, the volume of water produced being only $\frac{1}{2000}$ part of that of the gases.

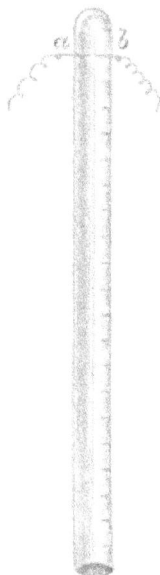

FIG. 151.

Another chemical effect of electrical discharge is the generation of **ozone**. It may be recognised when an electrical machine is worked, by its peculiar odour, and by its power of liberating iodine from a mixture of starch and potassium iodide.

(4) Mechanical effects.— Mention has been made of the mechanical effects of the disruptive charge in piercing the glass of a Leyden jar.

Exp. 184. Take a block of shellac through which a wire has been passed, having one end cut off flush with the surface, and the other so terminated that it can readily be attached to a chain or wire in connection with the outer coating of a Leyden battery. Lay a thin sheet of glass on the upper surface of the shellac. Insulate another wire, pointed at one end and terminated with a knob at the other, and then place it upright so that the point is exactly opposite the wire in the shellac cake. Attach a chain from the outer coating of a Leyden battery to the termination of the wire which passes through the shellac, and then charge the battery. By means of discharging tongs connect the rod of the upright wire with one of the knobs of the battery. A spark passes, and the glass will be pierced.

The apparatus shown in fig. 152 is arranged for this purpose.

B is a pointed conductor, supported by an insulated cross-piece.
The glass A is placed on an insulator, through which passes

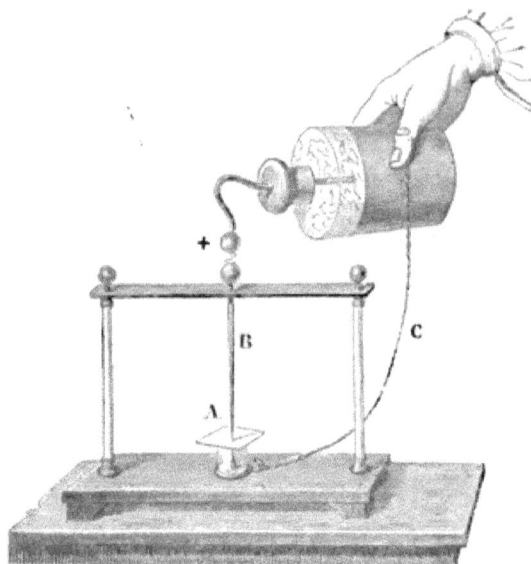

FIG. 152.

another pointed conductor in connection with the outer coating of
a Leyden jar or battery.   When the knob connected with the inner
coating is brought near the knob of B, the discharge passes, and
the glass is pierced.

Exp. 185. Place a sheet of cardboard on the shellac cake,
instead of the glass in Exp. 184.   1. Let the two wires be exactly
opposite each other, and, after the discharge, notice that the perfora-
tion is slightly frayed on both sides of the sheet, as though it were
pierced from the middle outwards.   2. Let one point be a little distance from the right or left of the other.   Notice that the hole is nearest
the negatively charged point.   This is known as Lullin's experiment.

Exp. 186. The last experiment may be varied by placing a card
or a piece of paper on the knob of a charged Leyden jar.   Discharge
the jar by placing one knob of the discharging-tongs on the outer
coating of the jar, and the other knob just above the cardboard.

(5) **Magnetic effects.—Exp. 187.** Make a spiral of silk- or
guttàpercha-covered copper wire.   Place a steel knitting-needle in
the interior.   Connect one end of the spiral with the outer coating
of a charged Leyden jar, and bring the other near the knob.   Pass

several discharges through the coil and prove that the needle has acquired magnetic properties. Fig. 153 represents a suitable arrangement for this experiment.

Exp. 188. A similar effect may be shown by placing a steel knitting-needle across a strip of tinfoil lying on a sheet of glass. Connect one end by means of a wire with the prime conductor of a machine, and the other with the earth (fig. 154). On working the machine, electricity passes from the prime conductor through the tinfoil. After a time the needle will become

FIG. 153.

magnetised. We therefore learn that a discharge of frictional electricity has a magnetising effect, but it must be remembered that a stationary charge produces no such effect.

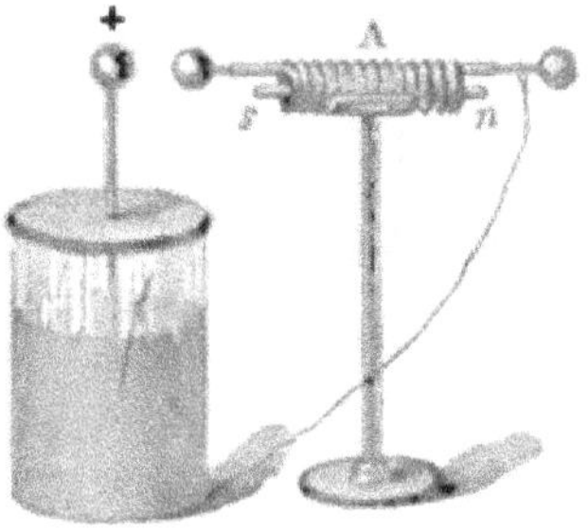

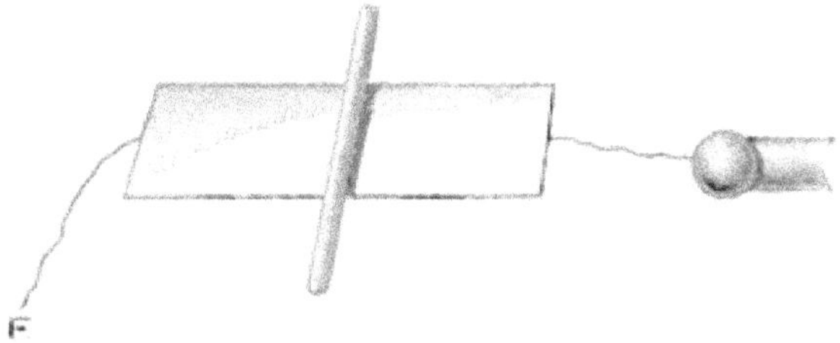

FIG. 154.

(6) Physiological effect.— Experiments illustrating this effect have already been described ; see Exps. 135, 153, and 163.

Note.—Extreme care must be taken with a large Leyden jar or battery, as the shocks may prove dangerous.

(7) Effects of flames and hot-air currents.—Exp. 189. Rub a vulcanite rod with flannel. Show that it is electrified by bringing it near an uncharged gold-leaf electroscope. Pass the rod through a Bunsen or spirit-lamp flame. Test as before. The rod has been completely discharged.

Exp. 190. Charge a gold-leaf electroscope, either positively or negatively. Hold a lighted taper above, but not in contact with, the disc. Observe that the leaves collapse ; the electroscope is, therefore, discharged.

Exp. 191. Repeat the last experiment, but cause the discharge by lighted cotton-wool soaked in ether or alcohol.

Exp. 192. Charge a gold-leaf electroscope. Place a spirit-lamp on the disc. There is no effect on the leaves. Light the lamp, and notice that the leaves immediately collapse, fig. 155.

Exp. 193. Raise the temperature of an iron ball to a red heat.

(*a*) Bring it above a positively charged electroscope. The electroscope is not discharged. As, however, the temperature sinks, it will be discharged.

(*b*) Bring the red-hot ball above a negatively charged electroscope. It is at once discharged.

We learn, then, that flames and hot-air currents are very good conductors of electricity; in fact, they discharge a conductor better than metallic points.

Duration of electric spark.—Exp. 194. Make a disc of white cardboard and paint a number of black sectors upon it, as shown in fig. 156.

FIG. 155.

Rotate this rapidly in ordinary light (a humming-top is useful for this purpose); the appearance is that of a disc with a grey surface. Darken the room and illuminate it with a spark from a Leyden jar. As the discharge takes place, notice that the disc appears to be perfectly still.

This shows the exceedingly small duration of an electric spark. It appears to last longer than it really does, from the well-known fact

FIG. 156.

that a flash of light produces an impression on the retina of the eye which remains for some time after the flash is ended. This is commonly called the 'persistence of impressions on the retina.'

EXERCISE XXVII.

1. Describe the luminous effects known as the spark, brush, and glow discharges.

2. I soak a piece of blotting-paper with a solution of potassium iodide, and another piece with a solution of potassium iodide to which a small quantity of starch has been added. I hold each piece to the prime conductor of a machine in motion. Describe the appearance in each case.

3. Explain as fully as you can the formation of water from its constituent gases by means of electrical discharge.

4. Give a simple experiment to prove that the duration of an electric spark is exceedingly short.

CHAPTER XV.

THE early observers soon perceived **that** lightning and thunder were similar in their nature **to the light** and crackling of the **electric spark.** Franklin proved an exact similarity between the two discharges in (1) giving light, **(2) speed,** (3) noise, **(4) conduction** by metals and moisture, **(5)** killing **animals, (6) fusing** metals, (7) rending bad conductors, **and** (8) **odour.**

He succeeded in drawing electricity from the **clouds** by means of a kite **having** a pointed wire attached **to it.** The kite was held by ordinary packthread, having a **key at** the end, to which was fastened **a piece of** silken cord **to** insulate **it** from the hand. At first he **was** unable to obtain **any** result, but a storm **of** rain came on, which, wetting **the** thread, made it a good conductor, and he then obtained sparks in sufficient quantity to charge a Leyden jar. Other investigators, with many different kinds of apparatus, have pursued this **subject.** (**The student** will find an account **of their investigations in** Ganot's ' Physics.')

From experiments of this kind it has been found **that** the atmosphere always contains electricity—usually positive, **but** sometimes negative. **During fine** weather the electricity is always positive, the **quantity** varying with the height above the ground. When **the** sky is clouded, **the** kind of electricity varies ; during rain, snow, **hail, the charge may be** at one time positive, and then change rapidly to negative.

Cause of Atmospheric Electricity.— Our real knowledge of **the** cause of atmospheric electricity is by no means large. It is, however, generally supposed to be derived from evaporation, which is continually going on from the surface of water upon the earth.

That electricity is derived from evaporation may be proved by the following experiment :—

Exp. 195.—Place a hot Hessian clay crucible on the disc of a gold-leaf electroscope (fig. 157.) Pour a solution of copper sulphate into the crucible. Rapid evaporation takes place, and the gold leaves diverge. Prove that the divergence is due to negative electrification.

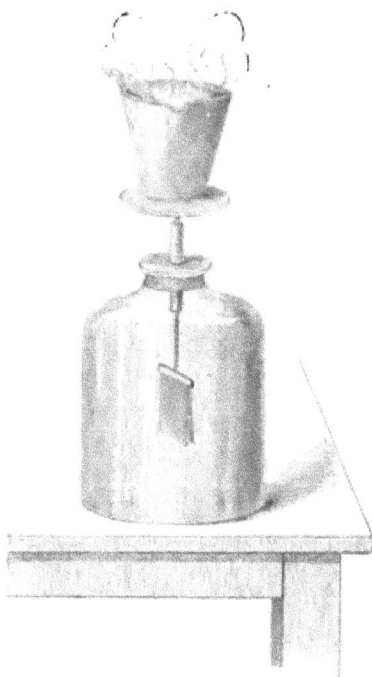

Fig. 157.

Water containing salts in solution becomes negative, the vapour becoming positive. In Exp. 195, however, the electrification is probably due to the friction between the particles of water and the sides of the vessel. How, then, is the vapour electrified positively in nature?

Some authorities suggest that the electricity of the atmosphere may be due to the friction of the particles of water against dry ice, both water and ice being present together in the higher regions. Upon this point, Professor O. J. Lodge, in one of his recent lectures, remarks : ' I can picture winds in the atmosphere driving the spray of mist against rock and ice surfaces, and so gradually producing a certain difference of potential between the upper layers of the atmosphere and the surface of the earth. I have spoken as if I thought the friction had to be between mist-globules and solid matter. It seems doubtful whether friction against air will suffice to render water electric.'

Whatever may be the origin of the electricity, it is certain that it is carried by the extremely minute water-particles. Each minute globule has a minute charge. **Now** suppose that eight of these globules, in falling, unite to form one globule ; the radius of the larger globule will **be** twice that of each of the eight, but the charge will be eight times as great.

Now $v = \dfrac{Q}{c}$ where v = potential

Q = quantity

c = capacity

now, the capacity of a sphere = the radius (p. 107), therefore, taking the quantity on minute globule as **1,** and the capacity as 1, we have

$$\text{potential} = \frac{1}{1} = 1,$$

and potential of globule formed by the eight minute globules

$$\text{coalescing} = \frac{8}{2} = 4.$$

Thus the potential of a cloud increases **enormously by the** coalescence of the minute particles, **and** as electrified clouds **act** by induction **on a** lower cloud or on the earth's surface, producing **an** opposite charge, the difference of potential between the clouds, or between **the** clouds and **the earth, increases so much that a discharge** takes place through **the air** between them. **These** discharges are flashes **of** lightning, **which** may be a mile or more in length.

Kinds of Lightning.—Three kinds of discharge have been distinguished :—

1. '**Forked lightning**' or the **zigzag flash.** No doubt this is due **to the** character of the air through which **the** flash passes, **certain** portions offering greater resistance than **others ;** in fact, its path is the **one of** least resistance.

2. **Sheet lightning** is probably **merely the illumination of** the cloud where the flash **occurs, or the reflection on** the **clouds** of a distant discharge. **It is called** *heat* or *summer* lightning **when the whole** horizon is illuminated **by** flashes so far distant **that the** thunder **is not heard.**

3. **Another,** and **very rare,** form is called **globular** lightning, **in which** *globes* *of fire* **travel** slowly, and then explode with **sudden violence.**

Thunder is the report which follows the discharge. It is probably due to the fact that the lightning heats the air in its path, producing sudden expansion and compression, which is followed by an extremely rapid rush of air into the rarefied space.

A thunder *clap* is produced when the path of the discharge is short and straight. The *rattle* is produced when the path is long and zigzag. *Rumbling* or *rolling* is produced by the echoes among the clouds.

When the discharge is near, the thunder immediately follows the flash ; if more distant, some time elapses between the instant at which the lightning is seen, and that at which the thunder is heard.

Light travels at a velocity of 190,000 miles per second, approximately; we may, therefore, consider that it is practically instantaneous. Sound travels, at the ordinary atmospheric temperature, with a velocity of 1,120 feet per second; the distance of the discharge can therefore be easily ascertained by multiplying 1,120 by the number of seconds which elapse between the lightning and the first sound of the thunder—e.g., suppose we hear the thunder 5 seconds after the flash is seen, the distance is 1,120 × 5 = 5,600 feet.

Return Shock.—Suppose that a thunder-cloud, many miles long, has a strong positive charge. By induction, the surface of the earth and the objects thereon, below the cloud, will be negatively charged. Let us further suppose that the difference of potential is so great that a discharge takes place between the cloud and the earth. The electrification of the cloud will be neutralised, not merely at the point of discharge, but even at points many miles distant. Imagine a person standing under the cloud, and therefore under its inductive influence, but situated at a place some distance from the point of discharge. When the discharge takes place, induction ceases, and the negative charge in the person is neutralised by the return of positive electricity from the earth. This sudden neutralisation is known as the **return** *shock*.

Lightning-conductors consist essentially of (1) a *rod*, usually of galvanised iron or of copper, elevated above the highest portion of a building, and terminated in a fine point ; (2) a *conductor*, between this rod and the ground, which often consists of a stout copper ribbon, or a rod of galvanised iron ; (3) the *earth-connection*, which is extremely important. The lower end of the conductor should terminate in several branches, and should pass into a well, or at any rate into moist earth. It is of the greatest importance that a lightning-conductor should be perfectly continuous.

The lightning-conductor illustrates (1) induction, (2) action of points. For, suppose a positively-charged cloud induces a negative charge on the surface of the earth below it. This negative electrification accumulates on, and is discharged from, the raised point, tending to neutralise the positive electricity in the cloud. Thus a lightning-conductor facilitates the discharge, but prevents disruption. Sometimes, however, the electricity accumulates to such an extent that the conductor is unable to *prevent* the disruptive discharge, and the lightning strikes. Even when this occurs, the conductor carries the discharge safely to the earth.

It has been calculated that the area protected by a lightning-conductor may be obtained thus :—

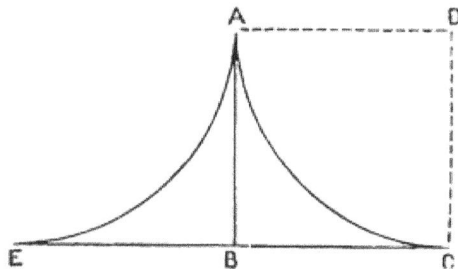

Fig. 158.

Let A B (fig. 158) be the height of the conductor above the ground. On A B construct a square, A B C D. With D as centre describe the arc A C. Similarly describe the arc A E on the other side. The protected area=the cone of which the vertical section is E A C.

An experiment to illustrate the necessity of having a continuous conductor may be performed with the apparatus shown in fig. 159. A represents the gable-end of a house, having a conductor, B, terminated in a knob at the top, and in a loop at the bottom. It is fitted with two small movable squares, C and D, having wires passing down the middle, so that when they are in the position

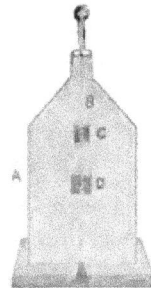

Fig. 159.

shown by the dotted lines in the diagram, the conductor is continuous. If, however, one or both are turned so that their wires are at right angles to the main conductor, the continuity is broken.

Exp. 196. (1) Attach a chain to the loop at the bottom of the conductor, and place the movable squares so that the conductor is continuous. Charge a Leyden jar in the ordinary way and hold the chain to the outer coating. On bringing the knob of the Leyden jar to that of the apparatus, the discharge passes through the wire without disturbing the squares.

(2) Place one of the squares so that the wire in it is at right angles to the main portions. Repeat the above experiment and notice that the square is forced out, owing to the interruption of the discharge.

The late Professor Clerk Maxwell suggested that a building should be covered with a network of metallic wires, so that the building would form, as it were, the interior of the conductor.

Dr. O. J. Lodge, along with a series of beautiful and striking experiments, has advanced statements, which, if accepted, will materially modify the old notions respecting the protection of buildings from lightning. He suggests that the conductor should consist of numbers of lengths of common telegraph-wire connected with large metallic masses such as leaden roofs. Balconies and other accessible places should not be connected. The connection must be thorough, and made at many points. The earth-connection should be deep enough to avoid damage to surface soil, foundations, gas and water mains. Barbed wire should be run round the eaves and ridges of the roof, so as to expose many points; but no rod should be run up above the highest point of the building.

The Aurora is a luminous phenomenon which occurs in the regions of the poles, and which depends upon the electrical state of the atmosphere. In the northern parts of the earth the phenomenon is known as aurora borealis, or 'northern lights'; while in the southern parts it is called aurora australis.

In the Arctic regions **the** aurora borealis occurs almost nightly, and it occasionally extends **over** very large areas. The light assumes various forms and colours ; **e.g.** (*a*) it sometimes appears merely as pale and flickering streaks, occasionally tinged with various **colours,** passing **from** the horizon towards the north magnetic pole ; (*b*) sometimes it forms **an arch ;** while (*c*) **at** other times it illuminates **the** whole sky.

Two chief facts point to the dependence of auroral displays upon the electrical **state of the atmosphere :—**

(1) Magnetic storms **(p. 40) always** accompany **them.**

(2) The rays converge **to** the **same point,** which **is a pro- longation of** the direction of a dipping-needle.

Many theories have been proposed to explain the phe- nomenon, but **Lemström showed** by his experiments in Lapland that it is due to currents of positive electricity passing from the higher regions of **the atmosphere to the earth.**

EXERCISE XXVIII.

1. How do you suppose thunder **and lightning to be produced.**

2. How do lightning-conductors **serve as a protection to buildings** on which they are placed ?

3. **A** thunder-cloud charged positively **comes over a** pointed lightning- conductor. The cloud gradually loses its charge of electricity by the action of the conductor. How is this accomplished ?

4. Describe **the** phenomenon known as ' the return shock.' **Give a** diagram.

5. **I have a** gold-leaf electroscope connected by means of a wire with **the earth. I place it near an** electrical machine in motion. The leaves **are unaffected. I take a** spark **from the prime** conductor ; **the** leaves diverge. Why ?

6. If a lightning-conductor **be placed on a house** having a leaden roof, **why** is it necessary to connect **the lead with the rod.**

7. Suppose **that a thunder-cloud is** overhead. **I have** a gold-leaf electroscope inside a room **of a house.** Will it be affected ? I take it outside the house. Will there **be any** alteration in the state of the electro- scope ? Give reasons for your **answer.**

8. **Find** the distance when **a discharge takes place if six** seconds elapse **between the light and** sound at the ordinary temperature of the air.

Ans. : 2,240 yards.

9. How long will the sound of thunder take to reach me if the discharge occurs 3,360 yards off?   *Ans.* : 9 seconds.

10. What will be the velocity of sound on a particular day if I hear the sound 6 seconds after I see a flash of lightning, the distance between the place of discharge and that of observation being 2,180 yards?

Ans. : 1,090 feet per second.   .

VOLTAIC ELECTRICITY.[1]

CHAPTER XVI.

THE SIMPLE CELL.

WE have frequently had occasion to notice that electricity spreads over the surface of conductors and flows along wires, threads of cotton, &c., from a point at a certain potential to one at a lower potential.

When electricity is supplied to a body as fast as it is removed, we have a continuous flow or *current* of electricity. Before discussing any method by which this can be done, it is necessary to perform a few preliminary experiments.

Fundamental experiments.—Exp. 197. Immerse a strip of commercial zinc (which is impure, owing to the presence of small particles of lead, iron, and other metals) in a vessel containing dilute sulphuric acid—one part of strong acid to twelve parts of water. Immediately it comes in contact with the acid, observe that a violent

---

[1] **Voltaic or** galvanic electricity is the term used to denote the phenomena caused by a current of electricity produced by chemical action, and is so called from the names Volta and Galvani. Galvani, a physician in Bologna, in 1786 noticed that when two dissimilar metals were in contact with each other, and the other ends were touching a nerve of the back and a muscle of the leg of a recently killed frog, a violent contraction occurred. He affirmed that the contraction was due to electricity generated by the animal itself. Volta, professor of physics in the university of Pavia, opposed this view, and asserted that it was due to the contact of two dissimilar metals. He upheld this opinion by a series of remarkable experiments. He first used his condensing electroscope and a compound bar of zinc and copper (see p. 140). He afterwards devised the voltaic pile (p. 180) and the 'crown of cups' (p. 179).

action begins ; bubbles of gas rise rapidly through the liquid from the surface of the zinc.

(*a*) Collect the gas by filling a test-tube with water and inverting it over the zinc. Now bring a lighted match to the mouth of the tube, and observe that the gas burns with a pale blue flame.   This is one of the tests for hydrogen.

(*b*) Evaporate a small quantity of the liquid, contained in a small porcelain evaporating dish, to dryness.   A white solid remains, which has been formed by the action of the sulphuric acid on the zinc.

(*c*) It can be seen that the zinc has worn away ; we may, however, actually ascertain the amount of loss by weighing the zinc before and after the experiment—e.g., in a particular experiment, a rectangular piece of zinc (5 inches × 2 inches) weighed 61·5 grammes; about half of which was immersed in the acid.   After the action had continued for half an hour, it was removed, washed, dried, and re-weighed. The weight was found to be 52·4 grammes; it had, therefore, lost 61·5 − 52·4 = 9·1 grammes, which had entered into combination with the sulphuric acid.

This experiment is an example of *chemical action.*[1]

The sulphuric acid acts upon the zinc, producing the white solid (sulphate of zinc) and a gas (hydrogen).

---

[1] When **chemical** action occurs, such a change is brought about in one or more substances that others are produced which differ from the original ones in properties.   There is not merely mechanical *mixing* of bodies, but there are produced entirely new and different substances.

It is advisable to point out to students unacquainted with chemistry that—

An *element* is a substance which cannot be split up into two or more different bodies.

A *compound* is a substance produced by the combination of two or more elements, and which can, therefore, be separated into its elements or into less complex compounds.

Symbols are abbreviations which represent elements in chemical writings, and which convey several important facts to a chemist's mind.

Formulæ are abbreviations for compounds.

Chemical reactions are expressed by means of *equations*, which are of the utmost value and must be carefully learnt.

e.g. Zn is the symbol for zinc, H_2SO_4 is the formula for sulphuric acid, and the action of these bodies on each other is represented by the equation :

$$Zn + H_2SO_4 = ZnSO_4 + H_2$$

(Zinc and sulphuric acid yield zinc sulphate and hydrogen.)

This reaction is expressed in chemical language by means of the following equation—

$$Zn + H_2SO_4 = ZnSO_4 + H_2$$

Exp. 198. Partially immerse a strip of copper in the vessel containing the dilute sulphuric acid. No evolution of gas takes place, and on examining the copper we find that there is no alteration in its appearance.

Exp. 199. Now remove the copper and immerse a strip of *pure* zinc. Again, there is no action, and no alteration in the zinc. It must be mentioned, however, that pure zinc is very difficult to obtain, so that a slight effervescence will probably occur when the strip is placed in the acid.

Exp. 200.—Amalgamate the zinc plate used in Exp. 197. This is done as follows : (1) Clean the zinc by dipping it into dilute sulphuric acid—of course in this instance the operation has been performed. (2) Cover its surface with mercury. To do this pour a little mercury upon the zinc, and immediately spread it over the surface by means of a piece of linen. The mercury unites with the zinc, forming a silvery-white amalgam of mercury and zinc.

Exp. 201.—Immerse the amalgamated zinc plate in the acid. Again no action takes place.

We learn from these experiments that dilute sulphuric acid does not attack copper, pure zinc, or amalgamated zinc.

The simple Voltaic Cell.—Exp. 202.—Place the strips of copper and amalgamated or pure zinc in the dilute acid.

(*a*) No action takes place when the metals are not in contact with each other.

(*b*) Allow the metals to touch *inside* the liquid. Notice that bubbles of gas begin to rise from the strip of *copper*, and that little or no action takes place on the zinc. If the gas be collected and tested as before, it will be found to be pure hydrogen.

A similar result occurs if

(*c*) direct contact of the metals be made *outside* the liquid, or if

FIG. 160.

(*d*) copper wires are attached to each plate, and their free ends brought in contact (fig. 160).

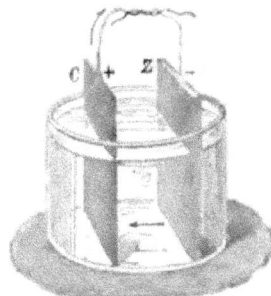

M

Voltaic Electricity

Such an arrangement is called a **simple cell**.

When the strips of metal are connected by means of copper wires, we say that the circuit is closed : before such connection is made the circuit is said to be open.

We thus learn that when two dissimilar metals are either in direct contact or in metallic connection with each other and with an exciting liquid, chemical action takes place—bubbles of gas generated at one plate being evolved from the other. We must now endeavour to grasp the theory of this action.

Chemical action in the cell.—The sulphuric acid (H_2SO_4) consists of molecules,[1] each molecule containing two atoms of hydrogen, one of sulphur, and four of oxygen. When the metals are in contact, the layer of molecules of the acid touching the zinc is broken up, by chemical action, into two groups

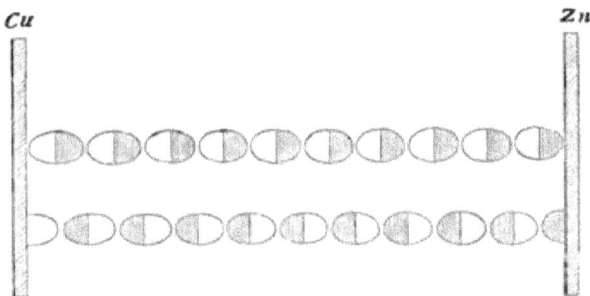

Cu   *Zn*

FIG. 161.

—H_2 and SO_4. The latter group cannot exist alone, and it, therefore, immediately combines with the zinc, forming zinc sulphate, and thus sets the hydrogen free. This nascent hydrogen (i.e. hydrogen at the instant of being evolved) possesses extremely active properties and is able to combine with the SO_4 in the next layer of molecules, forming H_2SO_4, and liberating its hydrogen. The third layer is again acted on in a similar manner, and so on, until finally the layer of molecules next the copper has its hydrogen liberated, which being unable to combine chemically with the copper, therefore

---

[1] **A molecule** is the smallest particle of a substance which can exist in a free state.

    An atom is the smallest particle of a substance which can enter into or be expelled from a chemical compound.

(1) coats its surface with bubbles, **many** of which

(2) rise through the liquid.

Fig. 161 will assist the student to understand this theory.

Suppose that each ellipse represents a molecule of H_2SO_4, **the** shaded half representing the group SO_4, and the unshaded half the group H_2.

The upper row shows the condition of the molecules before contact is made between the metals (Zn **and** Cu).

When contact is made, a rearrangement occurs amongst **the** groups, **as** shown in the bottom row. There is, as it were, an interchange of partners—the SO_4 **next** the zinc combining with **it to** form $ZnSO_4$; its H_2 combining with **the** next SO_4, and so on through **the** whole row of molecules, until finally H_2 appears **at** the surface of **the copper.**

EXERCISE XXIX.

1. Who was Volta? **Who was Galvani? Why do we hear** sometimes of voltaic electricity **and sometimes of galvanic electricity?** Is there any difference between them **?**

2. A plate of copper **and a** plate **of** amalgamated **zinc are** immersed in water rendered sour by sulphuric acid. Describe accurately and explain what is observed both before and after **the** plates are caused **to** touch each other.

3. **A piece of zinc** and a piece of copper are each carefully weighed. **They are then** connected by **a** copper wire, and dipped side by side into dilute sulphuric acid contained in an earthenware **jar.** After, say, half **an** hour, the pieces of zinc and copper **are** taken out of the acid, washed **and** dried, and weighed again. Would the weights be the same **as** at first **?** If not, how **and** why would **they differ?**

4. When copper and zinc are placed in contact in dilute sulphuric acid, hydrogen bubbles are given **off** from the copper, yet no bubbles **appear to** pass through the liquid. **How is this usually** explained **?**

Current of Electricity.—The chemical action between the liquid and the zinc, and the apparent transference of hydrogen to the copper described above is always accompanied by electrical action, i.e. there is immediately set up a *flow* or *current* [1]

---

[1] *Current* is a conventional term. **There** is no *flow* in the ordinary sense of the word. The expressions ' electricity flows,' ' current of electricity,' ' heat flows,' ' light travels,' convey certain ideas to our minds. What we actually know respecting a current of electricity is that the wire connecting the zinc and copper acquires new and different properties.

of electricity, when the two dissimilar metals are connected either by direct contact or by means of a wire.

We know that if two conductors are at different potentials, electricity will flow from the one whose potential is higher to that whose potential is lower : if, therefore, we show by actual experiment that the free end of the wire in connection with one plate gives a positive charge, and the free end of the other a negative charge, we conclude that a current will flow through the wire from the positively charged plate to the negatively charged one when they are connected.

Exp. 203.—Fasten one end of a silk-covered copper wire by means of binding screws to a plate of copper. Similarly fasten a copper wire to a plate of zinc. Dip the two metals into dilute sulphuric acid. Touch the bottom disc of a condensing electroscope (fig. 162) with the free end of the wire in connection with the copper, and the top disc with the free end of the wire in connection with the zinc. Remove the wires, and lift the top disc by means of the insulating handle. If the zinc and copper plates be large, and the electroscope be very

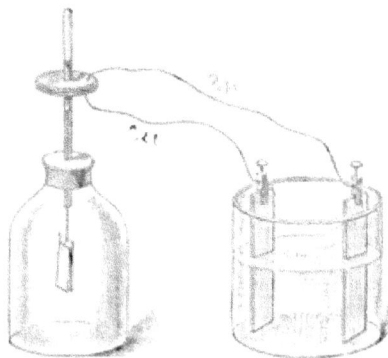

FIG. 162.

sensitive the gold leaves will diverge. Prove that the electrification in the leaves is positive.

Exp. 204.—Repeat the last experiment, but touch the lower disc with the wire from the zinc, and the upper one with the wire from the copper. Show that the gold leaves have a negative charge.

It is, however, extremely difficult to obtain a satisfactory result with a simple cell. To show the existence of the difference of potentials properly, we must use an apparatus somewhat in advance of our present knowledge, viz., a voltaic battery. An excellent one for this purpose is a 3 or 4-cell Grove's battery (described on pp. 174, 179). The wire from the platinum end is brought in contact with the lower disc, and the wire from the zinc end in contact with the upper one. The wires are then removed, and the upper plate lifted from the lower. The leaves diverge **widely with** positive electricity. The reason is that the wire in **connection** with the copper or platinum becomes positive, and therefore the disc upon which it is placed acquires a small positive charge ; the wire in connection with the zinc becomes negative, and therefore the disc in contact with it is negatively charged. When the upper plate is lifted, the capacity of the condenser diminishes very considerably, so that the small charge on the lower disc raises its potential so much that **the** gold leaves diverge. The last statement **will be** best understood from the equation, $v = \dfrac{Q}{C}$

(p. **127**) ; **when the denominator c of the** fraction $\dfrac{Q}{C}$ diminishes, the fraction increases, **and** therefore its equivalent, v, increases.

We learn **from** Exps. **203 and** 204 that positive electricity accumulates **at the end of the wire in** connection with **the** copper, and negative on the wire in connection with the zinc. A current will, therefore, flow through the connecting wire from the copper (which is called the *positive pole*) to the zinc (which is called the *negative pole*) when they are connected. And, as will **be** shown in Exp. 221, the current flows from the zinc to the copper through the liquid.

The starting-place of the current is, **however, at the** point where **the** zinc is acted on by the acid ; **the** waste of the zinc, no doubt, producing the power which drives the current through **the** liquid **to the** copper plate, and thence through the wire. **The** direction **of the current is** indicated by the arrows in fig. **160.** It ought to **be mentioned that it is a matter** of convention as to whether **we consider positive** electricity to pass through the wire from the copper to the zinc, or negative electricity from **the** zinc to the copper. Some authorities suppose that both currents flow. It is convenient to speak of the positive current as *the* current.

Exercise XXX.

1. State fully the meaning of the terms 'frictional electricity' and 'voltaic electricity,' and why they are employed.

2. What do you understand by an electric current?

3. Describe and explain any workable experiment to prove that the terminals of a galvanic battery differ electrically in the same way as the conductor and rubber of an electrical machine at work, but to a less extent.

4. You have a glass jar containing dilute sulphuric acid, into which a piece of zinc and a piece of copper dip without touching each other. How would you connect the zinc and the copper by means of wires with a poker, so as to make an electrical current pass through the poker from the handle to the point? Give a drawing.

5. Two copper wires connected, one with the zinc end and the other with the platinum end of a voltaic battery, but not connected with each other, are brought near a piece of sealing-wax that has been rubbed with flannel, and then nicely balanced on a point. Would the wires differ in any way in their action upon the sealing-wax? If so, how and why?

6. An insulated positively-electrified gold leaf hangs halfway between two vertical, insulated, and unelectrified copper plates, A and B, to each of which a copper wire is attached. If plates of copper and zinc, which are partially immersed in dilute acid, are touched by the wires attached to A and B respectively, and if afterwards the wires are dipped into the acid, describe the movements in each case.

Effects produced by the current.—(1) A magnetic needle, suspended on a vertical pivot, will be deflected if it be held near a wire through which a current is flowing.

(2) The temperature of the wire rises.

(3) Certain compounds are decomposed, if the ends of the wires from the two plates be dipped into them.

(4) If the wire be wound round a rod of soft iron, the iron becomes a temporary magnet.

(5) The wire itself becomes a magnet.

(6) It may also be mentioned that a peculiar taste is perceived when the wires from the poles of a cell are placed on the tongue. Sulzer noticed a similar effect in 1752, although it was not then known to be related to electricity. He placed a piece of lead and a piece of silver on the tongue, and when they were brought in contact, he perceived a 'certain taste resembling that of green vitriol.'

Exp. 205.—This effect can easily be repeated by placing a piece of zinc above, and a silver coin below the tongue. On bringing the edges of the metals in contact the peculiar taste will be at once perceived.

For the present a discussion of the above effects will be omitted, merely making use of the first one to indicate the existence, direction, and strength (the latter roughly) of a current.

Exp. 206.—Take a magnetic needle suspended so as to move freely in a horizontal plane. Bring the wire connecting the copper and zinc strips of a simple cell *above* and parallel to the needle. Observe that

(*a*) if the current pass from north to south, the N-seeking pole of the needle is deflected towards the east ;

(*b*) if from south to north, the N-seeking pole of the needle is deflected towards the west (fig. 163).

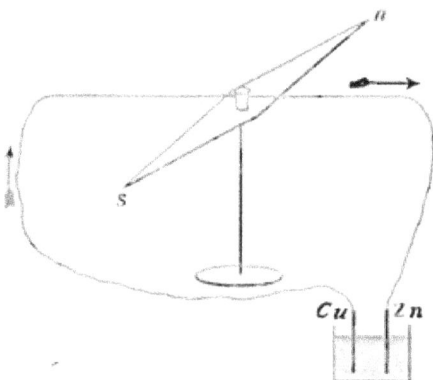

Carefully bear these results in mind.

FIG. 163.

The Galvanoscope.—When the current is weak and passes through a single wire it may not be able to deflect the needle, so that the construction of a more sensitive apparatus will be described.

It may be made very cheaply as follows :—

Make a wooden framework (5 or 6 inches long, $1\frac{1}{2}$ inches wide, and $1\frac{1}{2}$ inches deep). An ordinary cigar-box cut approximately to these dimensions answers well. It is advisable to glue rectangular blocks of wood, in the position shown in fig. 164, to support the

FIG. 164.

sides. By means of a chisel make a groove about an inch wide underneath the bottom of the frame.

Take silk- or guttapercha-covered wire and wind it round the frame ten or twelve times.

Fasten the framework to a rectangular wooden base, A (fig. 165), having first made a groove in it to correspond to the one underneath the bottom of the frame. Attach the ends of the wires to two binding-screws, B and C, fixed to the base. Now

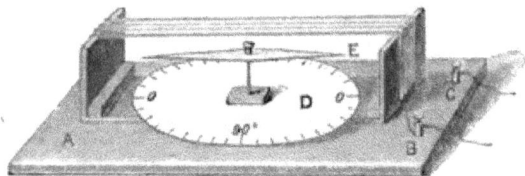

FIG. 165.

glue a circular graduated card, D, to the bottom of the frame, taking care that the zero of the scale is under the centre of the wires. Fix the eye of an ordinary sewing-needle into a rectangular piece of wood, and then glue the wood to the centre of the card. Place a magnetic needle, E, on the point of the vertical needle.

Such an apparatus is called a **galvanoscope**, and is as useful in voltaic electricity as a gold-leaf electroscope is in frictional electricity.

The principle of the action of the galvanoscope will be explained later on.

The instrument will be used as a **current indicator**, because

(*a*) The deflection of the needle shows the *presence* of the current ;

(*b*) The direction of deflection shows the *direction* in which the current is flowing ;

(*c*) The amount of deflection roughly indicates the *strength* of the current.

The apparatus will, however, be of no value to indicate very weak currents.

Exp. 207.—Attach the wires from a simple cell containing copper and zinc plates to the binding-screws, and observe that the effect is much greater than that obtained in the last experiment.

Electromotive series.—Exp. 208.—Replace the copper by strips of platinum, silver (half-a-crown answers well), iron, and a rod of carbon. Attach the wires to binding-screws as before, and observe that, in each case, the deflection of the N-seeking pole of the needle is in the same direction as when the copper strip was used. We, therefore, conclude that the current passes from these substances to the zinc through the wire.

It should also be noticed that the zinc has wasted away, and that the platinum, silver, iron, and carbon have undergone no such change.

Exp. 209.—Test the direction of the current, by means of the galvanoscope, when a plate of iron and a rod of carbon are partially immersed in dilute sulphuric acid. Notice that the current flows through the wire from the carbon to the iron.

Exp. 210.—Repeat the last experiment, using a plate of iron and a silver coin. The current flows through the wire from the silver to the iron.

From experiments similar to these the following substances, when partially immersed in dilute sulphuric acid, have been arranged so that, any two being taken, the current flows through the connecting wire from the *latter* to the former.

1. Zinc	6. Nickel	10. Silver
2. Cadmium	7. Bismuth	11. Gold
3. Tin	8. Antimony	12. Platinum
4. Lead	9. Copper	13. Graphite (carbon) [1]
5. Iron		

Thus taking zinc and iron, the current flows through the wire from the iron to the zinc; taking iron and graphite, it passes from the graphite to the iron.

The greater the distance on the list between any two substances, the greater the difference of potential.

The position of any substance on the above list varies considerably with (a) its condition, (b) the strength of the liquid, and (c) the nature of the liquid.

Any two being taken, the one standing first is said to be *electro-positive* to the second, which is *electro-negative*. It should be noticed that the electro-positive plate is the one acted on most by the liquid.

The student must not confound the terms positive and negative *plates* with positive and negative *poles*. For as the

[1] The order of the substances is that given in Ganot's *Physics*.

starting-place of the current is the part of the electro-positive *plate* (the zinc in a copper-and-zinc cell) which is immersed in the liquid, that part becomes positive, and we have proved by experiment that the part outside becomes negative. Now the parts of the plate outside the liquid are the *poles* of the cell ; thus we learn that the negative pole is that portion of the positive plate outside the liquid, and similarly the positive pole is that portion of the negative plate outside the liquid.

Local action.—We have seen that if contact be made between strips of copper and of pure zinc (or of commercial zinc which has been thoroughly amalgamated), when they are immersed in dilute sulphuric acid, no action, or at any rate a comparatively small one, takes place from the surface of the zinc (Exp. 202). If, however, the zinc be impure and unamalgamated, hydrogen gas rises not only from the copper plate but also from the zinc plate. This action causes the zinc to continually waste away without doing any useful work. This waste is due to what is called **local action**, which is explained as follows :—

(*a*) The impurities in the zinc consist of minute particles of other substances, e.g. iron, lead, arsenic, &c. Now suppose that a particle of iron is on the surface of the zinc, both being in contact with the acid, then, as shown in Exp. 208, a current is set up from the iron to the zinc *through the plate,* and from the zinc to the iron through the liquid. Other particles act in a similar manner, not only with respect to the zinc but with respect to each other, producing a large number of local currents, which cause continual waste of the zinc and diminution of the zinc-to-copper current.

(*b*) Another great cause of local action is due to the fact that various parts of a zinc plate, even if it be chemically pure, are often at different degrees of hardness. Now, the same metal having parts of unequal hardness, acts like two dissimilar metals in producing currents.

Theory of amalgamation.—We are now in a position to understand why commercial zinc is amalgamated (as in Exp. 200). It is to abolish local action, and so prevent the useless waste of the zinc.

(*a*) The zinc is dissolved by the mercury, forming an amalgam of mercury and zinc. The mercury, on the other hand, is incapable of dissolving the particles of iron ; at first they are therefore covered up, but as the zinc dissolves away they come to the surface, where they are quickly removed by any hydrogen bubbles which may be formed.

(*b*) The zinc amalgam forms a uniformly soft layer, so that the currents, which would be due to the varying hardness of the zinc plate, entirely disappear.

Polarisation of the Copper Plate in a Simple Cell.

Exp. 211. Fit up a simple cell with copper and zinc plates and very dilute acid. By means of a galvanoscope show that a current is flowing, and notice the amount of deflection on the graduated scale. Allow the action in the cell to continue for some time without disturbing the plates. Now observe that the amount of deflection has become smaller, showing that the strength of the current has diminished.

This falling off of the current is mainly due to the formation of a film of hydrogen bubbles on the surface of the copper plate. This formation of a film of hydrogen is called polarisation, and tends to weaken the current in two ways.

(1) The bubbles of gas, being bad conductors, offer great resistance to the passage of the current ; and the student will learn, when we treat of Ohm's law (p. 184), that the increased resistance involves a weakening of the current.

(2) Hydrogen gas is electro-positive ; thus the surface of the electro-negative plate becomes coated with an electro-positive substance, which tends to set up a current in the opposite direction.

Exp. 212. When a small deflection only is shown in the last experiment, carefully remove the copper plate and brush off the hydrogen bubbles. Immerse the plate again, and observe that there is a greater deflection, i.e., the current is stronger.

Thus the polarisation is, in a measure, prevented by continually brushing away the bubbles of gas. This, however, is an unsatisfactory method. It is best obviated by either

(1) Making the surface of the negative plate rough—as in Smee's cell (p. 176)—or

(2) Using a second liquid or substance which will act chemically upon the hydrogen. For this purpose the substance employed must necessarily be capable of oxidising and so destroying the nascent hydrogen. As examples we may mention nitric acid, bichromate of potash, and peroxide of manganese.

Exercise XXXI.

1. You are required to generate a voltaic current. How would you do it? How would you prove that you have really succeeded in producing a current.

2. What flows in the voltaic current—anything you can see, feel, taste, or smell? If not, what proof have you of its existence?

3. If you have within reach two wires, one to one end and the other to the other end of a voltaic cell, which is hidden, say how you could tell which wire is connected to the zinc end of the cell?

4. I place first a rod of carbon and a plate of zinc, and then plates of silver and zinc, in dilute sulphuric acid. I test the amount of deflection in each case by means of a galvanoscope. Will there be any difference? Give reasons.

5. Amalgamated zinc plates are employed in voltaic cells. How are the plates amalgamated, and what useful services does the amalgamation serve?

6. I have at my disposal strong sulphuric acid, water, a glass vessel, carbon, zinc, copper, and the necessary binding screws and wires. What must I do (*a*) to get the strongest current, (*b*) the weakest current, with these substances?

7. A steel knife and a silver fork are plunged into the same orange. What occurs when the metallic portions of both are united?

8. What is polarisation? How do you account for it? Explain how it may be prevented.

CHAPTER XVII.

COMMON VOLTAIC CELLS AND BATTERIES

Daniell's cell, or 'element' (fig. 166), consists of a cylindrical outer vessel, A, of copper, which forms the negative plate. Within this is placed a cylindrical porous cell, B (made of unglazed earthenware), and inside this a rod of amalgamated zinc, C. The liquid in the porous cell, and in contact with the zinc, is dilute sulphuric acid; that in the outer cell is a strong solution of copper sulphate; crystals of the same substance are placed on a perforated shelf, D, with which the outer cell is generally provided. These crystals of copper sulphate dissolve, so as to replace the copper which is removed by the chemical action taking place in the cell.

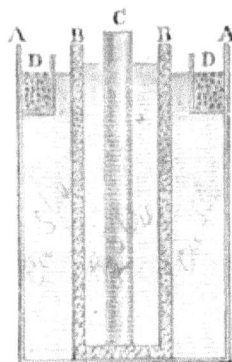

FIG. 166.

Fig. 167 represents a horizontal section of a Daniell's cell.

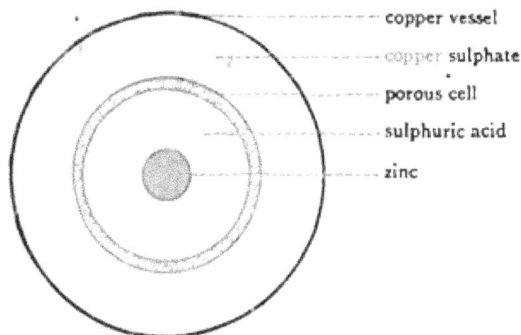

copper vessel
copper sulphate
porous cell
sulphuric acid
zinc

FIG. 167.

Exp. 213. Fit up a Daniell's cell as indicated above. Fasten copper wires, by means of binding-screws, to the copper vessel and zinc rod.[1]

Show that a current is passing by means of a galvanoscope. Allow the action to continue for some time. The amount of deflection of the needle remains the same ; the current is therefore constant.

Theory of the action of a Daniell's cell.—Fig. 168 illustrates the chemical actions which take place in the cell.

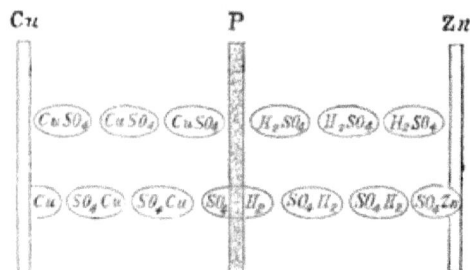

FIG. 168.

The top row indicates the molecules while the circuit is open. The bottom row shows their rearrangement after the circuit is closed.

The zinc is dissolved by the sulphuric acid, forming zinc sulphate ($ZnSO_4$) and liberating the hydrogen, which acts in a manner similar to that described on p. 162. When the liberated hydrogen reaches the porous cell, P, it comes in contact with the solution of copper sulphate ($CuSO_4$), which it decomposes, combining with the group SO_4, and liberating the Cu ; this immediately combines with the SO_4 in the next molecule, and so on, until eventually pure copper is deposited on the copper plate. By means of chemical equations the reactions are represented thus :

$$Zn + H_2SO_4 = ZnSO_4 + H_2.$$

The nascent hydrogen then acts on the copper sulphate, and we have :

$$H_2 + CuSO_4 = H_2SO_4 + Cu.$$

We thus learn how a second liquid destroys the hydrogen bubbles, and so prevents polarisation.

Grove's cell.—The outer vessel often consists of a flat cell

---

[1] In fitting up a cell care must be taken that good contact is made between the metals and the wires. To facilitate this the surface of the plates where the binding screws are attached, the screws themselves, and the ends of the wires, must be thoroughly cleaned by scraping them with a knife or by rubbing them with emery paper.

of glass or glazed earthenware, A (fig. 169) ; in this is placed a
strip of amalgamated zinc, B, bent round a porous cell, C, in
which is placed a piece of platinum foil, D.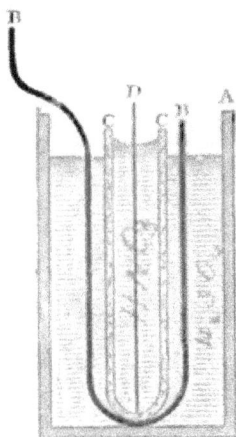
The liquid in the outer vessel, and there-
fore in contact with the zinc, is dilute
sulphuric acid ; that in the porous cell
is strong nitric acid.

The chemical actions in the cells are
represented by the following equations :

$$Zn + H_2SO_4 = ZnSO_4 + H_2.$$

When the nascent hydrogen reaches the
porous cell, it acts upon the nitric acid,
forming water and nitric peroxide gas,
thus—

$$H_2 + 2HNO_3 = 2H_2O + 2NO_2.$$

The nitric peroxide is liberated as a dark

FIG. 169.

red gas, which, unlike hydrogen, is unable to produce a film on
the platinum plate.

A Grove's cell it very convenient and powerful ; the platinum plates, however, cause it to be expensive, and the red
fumes of nitric peroxide are injurious.

Bunsen's cell is similar to a Grove's in principle and action ;
the only difference in construction is, that a rod of carbon re-

FIG. 170.

places the platinum plate. **Fig. 170** will sufficiently explain its
construction.

F is a cylindrical outer vessel of glass or glazed earthenware.

z is a zinc cylinder.

v is a porous cell.

c is a rod of carbon.

Dilute sulphuric acid is placed in the outer vessel, and strong nitric acid in the porous cell.

p shows these parts fitted up to form the complete cell.

Smee's cell (fig. 171) consists of a sheet of platinum or silver between two zinc plates, AA, dipping into a vessel containing

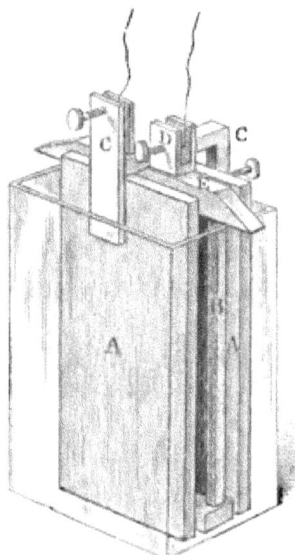 

FIG. 171.   FIG. 172.

dilute sulphuric acid. The platinum or silver is supported by a wooden framework, B, attached to a cross-piece, E, and is covered with finely divided platinum, the rough surface of which freely gives off the hydrogen bubbles. The zinc plates are fastened together by a binding-screw, c. The binding screw D is in metallic connection with the platinum or silver plate. The cell is, however, by no means constant, as the current falls off considerably after a few minutes.

The Bichromate cell is a very convenient form of cell for ordinary experimental work. It generally consists of a bottle of the shape shown in fig. 172, containing a zinc plate, z, attached to a brass rod, which moves up and down in a brass tube passing through an ebonite cover. By this means the zinc plate may be removed from the liquid when not in use. Two carbon plates, c, one on each side of the zinc, are attached to the cover. The liquid is a mixture of sulphuric acid (H_2SO_4) and potassium bichromate ($K_2Cr_2O_7$). The proportions vary, but they often consist of H_2SO_4, 2 ozs. by weight; $K_2Cr_2O_7$, 2 ozs.; and water, one pint. The chemical action in this cell is too complicated for description in this work.

The difference of potential remains constant for a short time, and then rapidly decreases. The cell is chiefly of value when an intermittent current is required.

Leclanché's cell (fig. 173) consists of a rod of carbon, c, placed in a porous cell, P, in which are also placed small pieces of carbon, and powdered per-oxide of manganese. A piece of lead, L, is soldered to the top of the carbon, to which a binding-screw is fixed. A rod of zinc, z, and a solution of ammonium chloride are placed in the outer vessel G. The zinc dissolves in the solution of ammonium chloride (NH_4Cl), forming zinc chloride ($ZnCl_2$), which at once combines with the ammonia (NH_3), and liberating hydrogen, which is slowly oxidised by the peroxide of manganese. The current given by this cell is continuous for a few minutes only, owing to the formation

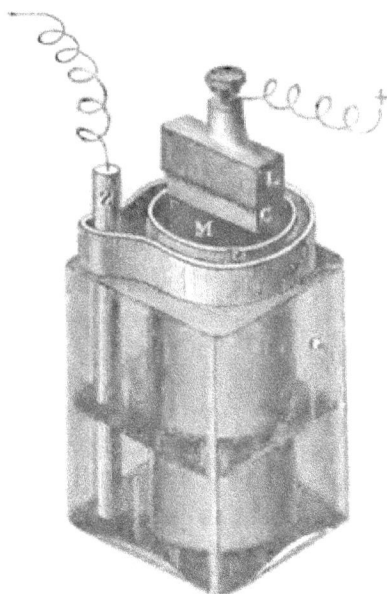

FIG. 173.

of hydrogen bubbles, which are unacted on by the peroxide of manganese. If, however, it be allowed to rest, it regains its original strength. It is thus well adapted for ringing electric bells and for occasional use in telegraphy.

Gravitation cells.—The use of porous vessels has several objections, so that cells have been devised in which the liquids

N

are separated merely by a difference in density. As an example the cell devised by Callaud may be mentioned (fig. 174). v is a glass or earthenware vessel, c is a copper plate soldered to a guttapercha-covered wire. z is a zinc cylinder. A layer of crystals of copper sulphate ($CuSO_4$) is placed on the copper plate, and the cell is then filled up with water. These crystals dissolve in the water, forming a solution which gradually diffuses, and so comes in contact with the zinc, forming zinc sulphate ($ZnSO_4$), which, being lighter than the $CuSO_4$, floats on it. The action is similar to that of a Daniell's cell, described on p. 174.

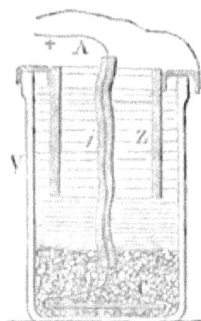

FIG 174.

Voltaic batteries.—Cells, sometimes called *pairs, couples,* or *elements,* may be grouped in several ways.

(*a*) *Simple circuit,* or *multiple arc* is that arrangement in

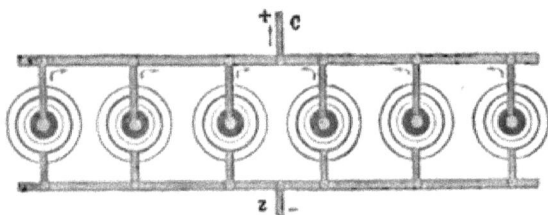

FIG. 175.

which all the zincs are connected with one another, and all the carbons [1] with one another (fig. 175).

(*b*) *Compound circuit,* or '*in series,*' is that arrangement in which the carbon of the first is connected with the zinc of the

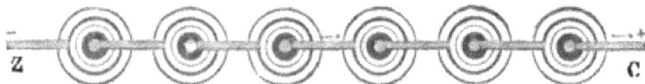

FIG. 176.

second ; the carbon of that with the zinc of the third, and so on (fig. 176).

(*c*) *A combination of* (*a*) *and* (*b*). In fig. 177 each set of

---

[1] This illustrates the arrangement with Bunsen's cells.

three cells is joined 'in series,' while the two at each end are joined in simple circuit. A group of cells arranged in any of these ways is called a **voltaic** or **galvanic battery**.

Fig. 178 represents 10 Bunsen's cells arranged in compound circuit. Each carbon is connected with the next zinc by means of clamps, *m n*, and a thin rod of copper, *c*, shown at the top of the figure.

The first zinc and the last carbon are provided with binding-screws, to which the wires are attached.

Grove's Battery (fig. 179) consists of a number of cells arranged in compound circuit, the zinc of one being connected with the platinum of the next by means of binding-screws, *m*. The free zinc at one end is connected by

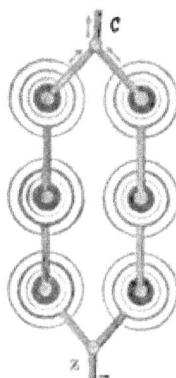

FIG. 177.

the binding-screw *a* to a copper wire. The free platinum is supported by a piece of brass attached to a box, B (in which the

FIG. 178.

cells are placed) by means of a binding-screw, *b*, which also serves to connect it with a copper wire.

Volta's Crown of Cups.—The earliest forms of battery were constructed by Volta. One of these, shown in fig. 180, he called the '*Couronne de Tasses*' or *Crown of Cups*. It con-

N 2

sisted of a number of glass cups partly filled with brine or dilute sulphuric acid, in each of which two strips —one of zinc

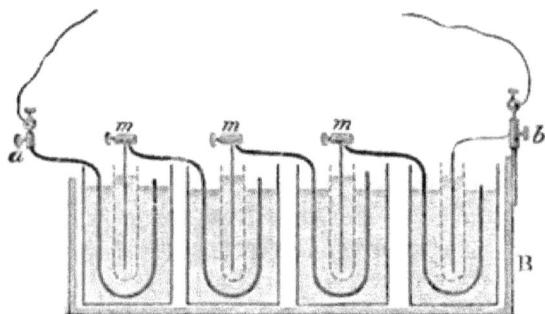

FIG. 179.

and the other of copper—were placed. The metals were connected ' in series.'

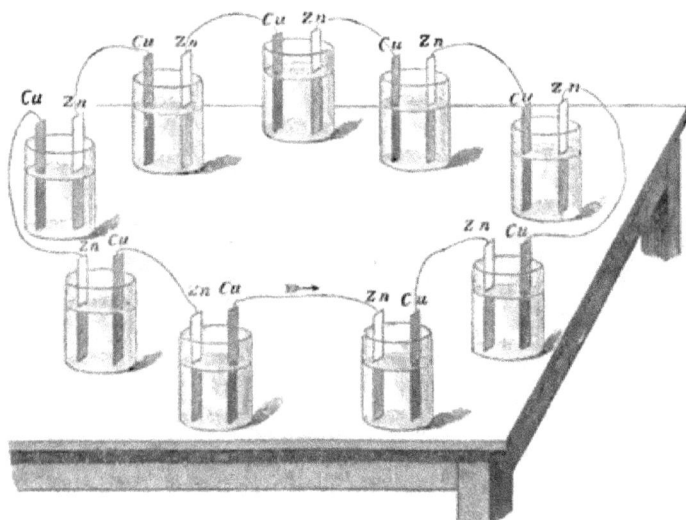

FIG. 180.

The Voltaic Pile (fig. 181) consists of a large number of discs of copper and zinc, often soldered together, with a layer of flannel (saturated with brine or dilute sulphuric acid) between each pair. The discs are kept in the same order—copper, zinc, flannel; copper, zinc, flannel, and so on. The column is placed

in a wooden frame. When the bottom copper is connected
with the top zinc by means of a wire, a current passes in the
direction indicated by the arrow. At
first sight the direction appears as
though it were contrary to that given
on p. 169, but the fact is the zinc at the
top may be considered as a metallic
extension of the copper next to it, and
the copper at the bottom as an extension of the next zinc. The current
starts from the zinc, passes through the
exciting liquid in the flannel, and
thence to the copper.

 This apparatus was devised by
Volta to prove the *contact theory* of
electricity. As already shown on
p. 141 the mere contact of two dissimilar metals produces a difference of potential, but chemical action is, no doubt, the agency
which maintains this difference so as to produce a current.

 Zamboni's Dry Pile consists of a very large number of
paper discs, covered on one side with tinfoil and on the other
with powdered peroxide of manganese (MnO_2) piled together
in a glass tube in the order—tinfoil, paper, peroxide of manganese. With several thousands of these discs, electric sparks
are produced. The action of the pile is, no doubt, kept up
by the moisture in the atmosphere.

 Bohnenberger's Electroscope consists of a single gold leaf
suspended midway between two metal plates, which are connected with the two poles of a dry pile (the manganese peroxide
end of the pile becomes the positive pole and the zinc end
the negative). The kind of electrification of the body to be
tested will therefore be shown by the attraction of the leaf to
one pole, and repulsion by the other.

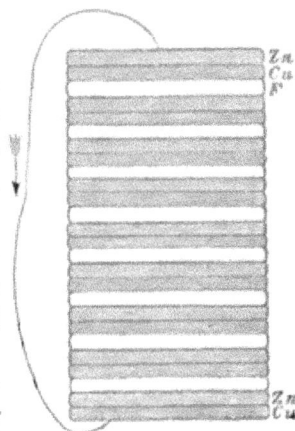

Fig. 181.

EXERCISE XXXII.

 1. What advantage has a two-fluid cell over a one-fluid cell?

 2. Sketch a Grove's battery. You are required to charge the battery.
How will you do it?

3. What is the part played by nitric acid in a Grove's battery?

4. What are the materials used in the construction of a Daniell's cell, and what chemical changes occur in the cell when in action?

5. Draw and describe the various parts of (1) Bunsen's and (2) a bichromate cell, and indicate in the sketch the direction of the current when the terminals are joined.

6. The platinum and copper plates of a Grove's and a Daniell's cell are connected by a wire. Would there be a current if the zinc plates were also connected, and, if so, in which direction would it flow? What reason have you for your answer?

7. If a charged battery is to be kept for some time ready for use, why is it important to take care that the ends of the battery are not connected outside the battery?

8. Make a sketch of three Grove's cells arranged (1) ' in series,' (2) in simple circuit.

9. Describe the chemical action which goes on in the voltaic cell itself. You must, of course, here describe the particular cell you have in view.

Electromotive force.— It was shown in Exp. 203 that there is a difference of potential between the two ends of the wires in connection with the copper and zinc plates of a simple cell.

That which produces this difference of potential is known as **electromotive force** (generally contracted to E.M.F.).

The term *force* applied to this agency is by no means appropriate, as a force is that which moves, or tends to move, *matter*, while E.M.F. is that which produces, or tends to produce, a transfer of. *electricity*.

We will now consider the relation between the difference of potential and the arrangement of cells in a battery.

(*a*) In the *simple circuit* arrangement, since the zincs are connected they form, as it were, one large zinc plate ; similarly the coppers (in a Daniell's battery) form one large copper plate.

Now Volta demonstrated that the difference of potential between two metals did not depend upon their size, but merely upon the kind of metals employed.

Therefore, in this method of grouping, the difference of potential is the same as that of a single cell.

(*b*) When similar cells are arranged *in series*, the difference of potential increases with the number of cells. Let us consider two cells, A and B.

There is a certain difference of potential between the zinc and copper of A, and an equal difference between the zinc and copper of B; but when the zinc of A and the copper of B are joined, their potentials are equalised; whence the difference of potential between two cells arranged in series is twice that of one.

As a general law the difference of potential, or, as we may consider it, the E.M.F. of a number of like cells thus arranged is equal to the E.M.F. of one cell multiplied by the number of cells.

Resistance.—The student must not fall into the error of supposing that when there is an increase of E.M.F. the current is necessarily stronger. Another factor has to be brought into consideration, viz. the *resistance* which the current has to overcome both in the battery itself and in the connecting wire.

Resistance in wires can be well illustrated by reference to the flow of water through pipes. Suppose that we have two vessels, A and B, containing water, so that the level in A is always higher than that in B; suppose further that they are connected at their bases by a pipe. We know that, in a given time, a certain quantity of water will flow through the pipe from A to B.

(1) Now suppose that the pipe is choked with some substance—e.g., sawdust—which obstructs the flow of the water, then, of course, the quantity flowing through the pipe in a given time will be less than in the case where the pipe was originally empty, the difference between the two quantities being due to the obstruction—i.e., the quantity flowing through in a given time depends upon the *resistance* offered by the pipe.

(2) If we have a pipe similar to that used in the previous case, and choked with sawdust in the same way, but four times as long, we then have four times the amount of obstruction, so that a quarter of the quantity will flow through the pipe in the given time.

(3) Again, if we place four pipes of the length used in the first case side by side, then the difference of level in A and B remaining constant, four times the quantity of water will flow

through the pipes in the given time—i.e., the water meets with one quarter of the resistance.

Similarly (*a*) the quantity of electricity which flows through a wire depends upon its resistance, the E.M.F. remaining constant—i.e., the greater the resistance, the less the flow.

(*b*) A long wire offers a greater resistance to the passage of electricity than a short one of the same material.

(*c*) A thick wire offers a less resistance than a thin one. Generally, if R = resistance, *l* = length of a wire, *a* = area of resistance of cross section.

$$\text{then } R = \frac{l}{a}$$

which is expressed in words thus :—The resistance in wires of the same material is directly proportional to the length, and inversely proportional to the area of the cross section.

Resistance of Liquids and Gases.—Liquids have, compared with metals, high resistances for current electricity ; and moreover different liquids have different resistances—e.g., in pure water it is very high, while in brine and dilute acids it is comparatively low.

The resistance of gases is high. Thus the hydrogen which accumulates on the negative plate of a cell increases the resistance so much that the current is considerably weakened.

Strength of current is defined as *the quantity of electricity which flows across any section of the circuit in one second.*

Ohm's Law.—The relation that exists between the strength of a current, the electromotive force, and the resistance, was given by G. S. Ohm, in the following law :—

The strength of a current varies directly as the electromotive force, and inversely as the resistance.[1]

---

[1] Electromotive force is measured in volts. Without much error we may consider the unit volt to be the E.M.F. of a Daniell's cell.

Resistance is measured in ohms. An ohm is the resistance of a column of mercury one square millimetre in section and 106 centimetres long at 0° C.

(A metre = 39·37 inches ; a centimetre = ·3937 inches ; 106 centimetres = nearly six times the length of a page of this book ; a millimetre = the thickness of 6 of these pages.)

Current strength is measured in ampères. An ampère is the current given by an E.M.F. of one volt through a resistance of one ohm.

It may be expressed by means of an equation, thus

$$c = \frac{E}{R} \quad \text{(i.)}$$

where c = current strength, E = electromotive force, R = resistance.

If we divide the whole resistance **R** into the internal resistance R′ and the external resistance *r*, then

$$c = \frac{E}{R + r} \quad \text{(ii.)}$$

We therefore learn that (1), the resistance remaining constant, the strength of the current increases as the electromotive force increases ; and (2), the E.M.F. remaining constant, the strength **of the** current increases as the resistance (whether in the battery or in the wire) diminishes.

Now, as previously mentioned, the E.M.F. depends upon the substances employed and upon their condition, and **not** upon the size of the plates. To increase the E.M.F. it is, therefore, necessary to join the cells 'in series.'

The resistance depends upon various conditions—e.g., the length and **thickness of** the wires, the size of **the** cells, the conducting power **of the** liquid, the distance **between** the plates, &c.

To diminish the *external* resistance **we must** have the wires, (1) short and (2) **thick.**

To diminish **the** *internal* resistance we may (1) **bring the** plates near together, (2) increase their size, (3) the cells must be grouped in simple circuit (p. 178).

Exercise XXXIII.

1. Explain why the external resistance in a battery is diminished by making the wires short and thick.

2. Explain why the internal resistance is diminished **by (1)** bringing the plates nearer together and (2) increasing their size.

3. Explain the meaning of the terms—electromotive force, resistance, current, strength.

4. What is meant by *ohm, ampère, volt*?

5. **What** would **be the** resistance at 0° C. of a column of mercury 154

centimetres long and two-fifths of a square millimetre in cross section (see foot-note, p. 184).　　　　　　　　　*Ans.* : 3·6 ohms nearly.

6. Find the strength of a current in a circuit with an E.M.F. of 10 volts and a resistance of 4 ohms?　　　　　*Ans.* : 2·5 ampères.

7. Find the resistance in a circuit if an E.M.F. of 12 volts gives a current of 1·5 ampères.　　　　　　　*Ans.* : 8 ohms.

8. Find the E.M.F. if the current strength be 4·5 ampères, and the resistance 12 ohms.　　　　　　　　*Ans.* : 54 volts.

CHAPTER XVIII.

THERMAL AND MAGNETIC EFFECTS OF A CURRENT.

Exp. 214. Pass a current from a 3- or 4-cell Grove's battery through a moderately thick copper wire, and notice that no appreciable heat is produced.

Exp. 215. Fasten a few inches of thin platinum wire to the copper wires attached to the battery. The wire speedily becomes red-hot, and then, if not too long, white-hot. The heat developed is sufficient to fuse very thin platinum wire.

We have previously learnt that thin wires offer great resistance to the flow of electricity, and we now learn that an increase of temperature accompanies an increase of resistance. The laws which relate to the production of heat by electricity have been investigated by means of the apparatus shown in fig. 182. A wide-mouthed bottle was inverted, with its stopper *b*, into a wooden box. Holes were bored in the stopper to admit two stout platinum wires, connected with the binding-screws, *s s*. The free ends were fitted with platinum cones, to which the wires, to be experimented upon, were fixed. The vessel contained alcohol, the temperature of which could be ascertained by means of the thermometer *f*.

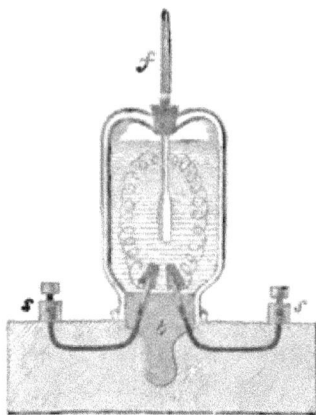

FIG. 182.

The following law, known as *Joule's law*, was established by means of an apparatus of this kind.

The number of units of heat [1] produced in a wire varies as (1) its resistance, (2) the square of the strength of the current, (3) the time that the current flows.

Exp. 216. Take the platinum wire of such a length that it merely becomes red-hot. When the wire is in this state bend it and dip the middle portion in a vessel of cold water. Observe that the ends immediately become white-hot.

In this experiment the cooling of the wire has diminished the resistance, so that more electricity flows in a given time. The increased flow meets with a great resistance in the uncooled portions, and, therefore, raises their temperature.

Exp. 217. Cut fine platinum wire (No. 33 B.W.G.) into six or eight pieces, each an inch long. Similarly cut fine silver wire into inch lengths. Bend a piece of the platinum wire into a loop, attach to this a piece of the silver wire also bent into a loop, and so on, forming a chain of alternate loops of platinum and silver. Place the chain in the circuit (using a 5-cell Grove's battery). Notice that the platinum links become white-hot, while the silver links remain comparatively cool.

This effect arises from the fact that silver is a good conductor, and therefore offers small resistance to the current-flow ; while platinum is a much worse conductor — its resistance being so great that the wire becomes white-hot.

This experiment has a pretty effect in a darkened room, as the platinum links become luminous, while the silver ones remain invisible.

With respect to the luminous effects of the current it may be mentioned that they have been utilised and developed for lighting purposes. The principles [2] will be understood from the following considerations.

The Voltaic Arc.—When two pencils of charcoal are connected with the terminals of a Grove's battery of 40 or 50 cells, or with any powerful generator of current electricity, and the points

---

[1] A unit of heat is the amount of heat required to raise 1 gramme of water from 0° C. to 1° C.

[2] The principles of electric lighting are *merely indicated* here. The student who wishes to study this subject must read a more advanced work.

of the pencils are brought in contact, and then separated by about ⅛ inch, a luminous *arc* is produced between them, which is of extreme brilliancy.

The voltaic arc is produced as follows :

(1) When the carbons are in contact, a current passes, which, meeting with great resistance at the points, makes them white-hot.

(2) When they are removed a short distance apart, the air between them is heated, and small white-hot particles pass from the positive to the negative pole. In consequence of this, the positive pole becomes hollow, and the negative pole pointed (fig. 183).

FIG. 183.

FIG. 184.

Incandescent Lamps.—For illuminating rooms, &c., incandescent lamps are used. Fig. 184 represents one of these, which is known as the Swan lamp. It consists of a glass bulb containing a loop of carbon which is raised to a white heat by means of a current. The bulb is exhausted and then hermetically sealed. The nature of the filament is of the greatest importance, and many suggestions have been made respecting it. Platinum fuses too readily. It commonly consists of a thread of carbon, which is prepared in many ways—e.g., Swan prepares it by immersing cotton in sulphuric acid, and then raising it to a high temperature in a closed vessel, to carbonise it. Edison uses carbonised bamboo.

Exercise XXXIV.

1. If the same current be sent through a silver wire and an iron wire of the same thickness and length, the iron wire will become hotter than the silver. Why?

2. A current flows through a copper wire, which is thicker at one end than at the other. If there is any difference either (1) in the strength of the current at, or (2) in the temperature of, the two ends of the wire, state how they differ from each other, and why?

3. How does the heating of a wire depend upon the resistance, the strength of the current, and the time the current lasts?

Magnetic effects of the current.—We have seen that if a magnetic needle be allowed to come to rest, it sets itself in the magnetic meridian, and that if a wire, through which a current is flowing, be brought *above* the needle, the latter is deflected so that it tends to set itself at right angles to the meridian. We shall now learn that a magnetic needle is deflected when the wire is placed in other positions.

The experiment by means of which Oersted showed the effect of a current on a magnetic needle is known as **Oersted's experiment.**

Exp. 218. Place the apparatus shown in fig. 185 in the magnetic meridian.

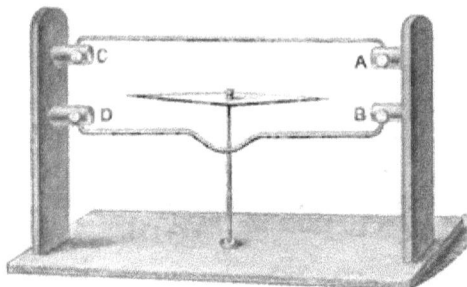

FIG. 185.

(a) Attach the wires from the terminals of a cell or battery to the binding - screws, A C. The current now passes *above* the needle. Observe that the needle is deflected in the direction given in the following table.

(b) Next attach them to the binding-screws B D ; the current now passes *below* the needle. Observe the direction of deflection.

Position of wire	Direction of current	Direction of deflection of N-seeking pole
above the needle	N. to S.	E.
,,	S. to N.	W.
below the needle	N. to S.	W.
,,	S. to N.	E.

Ampère's rule.—Ampère gave an exceedingly useful rule by which the direction taken by the **N**-seeking pole of a magnetic needle, when under the influence of a current, can be easily ascertained.

Let the observer imagine himself **to be** *swimming* **in** *the wire in the direction of the current, with his face turned towards the needle, then the N-seeking pole will* **turn** *in the direction of his* **LEFT** *hand.*

Exp. 219. Repeat the last experiment, **and notice the** direction of deflection by aid of the above rule.

Of course when the wire is *below* the needle, the observer must swim on his back. When using this rule, consider the motion of the *N-seeking pole merely*, not the direction of the *needle as a whole*.

Exp. 220. Hold the wire conveying the current in various

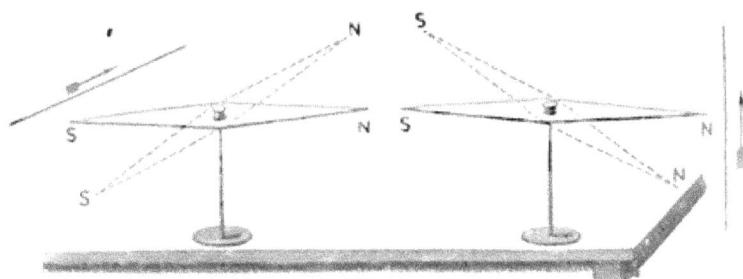

FIG. 186.

other positions, (fig. 186), and **observe** that Ampère's rule can still be applied.

Current in the cell.—**Exp. 221.** Connect the poles of a 5-cell Grove's battery by means of a wire. Take a magnetic needle balanced on a point so that it moves horizontally. Lift the box containing the cells and bring it over the needle. Observe that (1) when the platinum end of the battery lies towards the north, the needle is deflected so that the N-seeking pole turns towards the west ; (2) when the zinc end of the battery lies towards the north, the N-seeking pole of the needle is deflected towards the east.

Take care that the wire is bent so that it lies some distance above the battery, and has no influence upon the needle.

From this experiment we have, therefore, proved that the current is moving in the battery itself from the zinc to the platinum.

Multiplying Effect of Currents.—Exp. 222. Suspend a magnetised knitting-needle by unspun silk, so that it is able to move in a horizontal plane. Bend a copper wire into the shape of a rectangle. Attach the ends of the wire to a voltaic cell, and place the rectangle so that one wire is *above*, and the other *below* the needle, thus forming a circuit round the needle (fig.

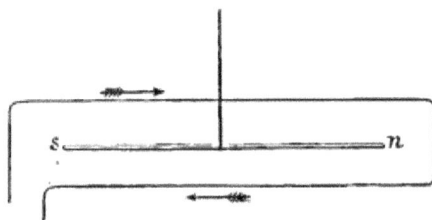

FIG. 187.

187). Observe that the needle is deflected. Let us examine the effect of this arrangement. Applying Ampère's rule we find that the current in *each* wire will urge the N-seeking pole of the needle in the same direction. Thus the effect is increased, for the current passes through four wires, which assist each other to produce a deflection in the same direction.

Exp. 223. Bend a wire into a *loop* and place it round the needle (fig. 188). An effect similar to that obtained in the last experiment will be produced.

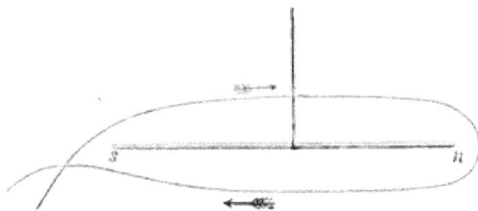

FIG. 188.

These experiments may also be made by means of the apparatus used in Exp. 218, if c d be attached to the wires from a cell, and a b be joined by a wire.

The student will easily understand why the effect of a current on a magnetic needle is materially increased by winding the wire many times round the needle, as, for example, in the galvanoscope. It must, however, be remembered that with the same cell or battery the number of circuits cannot be increased indefinitely, for the resistance increases as the length of wire becomes greater.

We have learnt (p. 50) that one of the methods of obviating the directive influence of the earth on a magnetic needle is by using an astatic pair. It is therefore evident that such a combination is more readily deflected by a current flowing through a wire, which is coiled round one of the needles.

Exp. 224. Bend a wire as shown in fig. 189 round the lower needle of an astatic pair. As before, attach the wire to the terminals of a voltaic cell. Observe the deflection.

The effect of the current along the upper and lower wires will be understood from the following explanation :—

The current along *no* and along *pq*, acting in opposite directions, tends to deflect the needle *a'b'* in opposite directions ; but the current in *no* is nearer the needle than that along *pq*, and therefore the former has a greater effect than the latter.

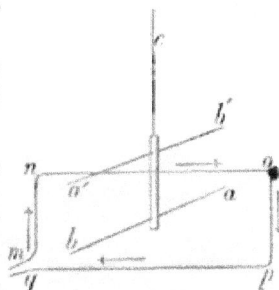

Fig. 189.

Now the current *no* causes the north-seeking pole *a'* to come out from the plane of the paper.

Again, the currents along *no* and *pq* acting on the needle *ab* will cause the N-seeking pole *a* to pass through the plane of the paper, and therefore, will cause *b* to come out from the plane of the paper —i.e. the effect of the currents on both needles will be to urge the opposite poles *a'*, *b*, in the same direction.

Astatic Galvanometer or Multiplier is an instrument similar to the galvanoscope in principle, but it is very much more delicate ; in fact, it is used to measure feeble currents only. It consists of an astatic pair, hung by a fibre of unspun silk, the lower needle lying within a coil of many turns of wire, which is wound upon a wooden frame. The two ends of the coil terminate in binding-screws, *i o*, fig. 190. Above the coil is a graduated circle, c, with a central slit cut parallel to the direction of the wires in the coil. The zero of the graduated circle lies at the end of the diameter passing through the axis of the slit,

o

and on each side degrees (up to 90°) are marked. A glass shade rests upon the plate P, the latter being provided with three levelling-screws.

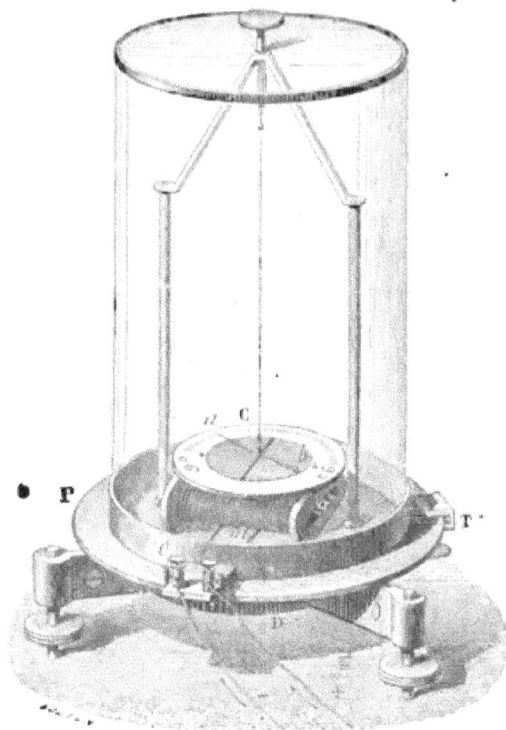

To use the instrument the frame carrying the coil is first placed parallel to the needles. Wires from the cell are then attached to the binding-screws *i o*.

The strength of two currents can be compared by the galvanometer, provided that the deflection of the needle caused by the two currents respectively is not greater than 10° to 15°, *for in the case of small angles the strength of a current is proportional to the angle of deflection* : i.e., if a current deflect the needle through 10° it has twice the strength of one which deflects it through 5°.

Fig. 190.

If the deflection be greater than 15°, it is possible to compare the strength of two currents by introducing a known resistance into the circuit, or by 'calibrating' the galvanometer, i.e. by finding the relative values of the deflection by actual experiment with a standard instrument.

Very often the instrument is used as a *galvanoscope*—to ascertain the presence and direction of currents—rather than as a *galvanometer*—to measure their strengths.

EXERCISE XXXV.

1. If you wanted to prove that an electric current acts upon a magnetic needle, give me a clear statement of how you would proceed.

2. You are required to prove that a current of electricity is passing through a telegraph wire to which you have access. How would you do it?

3. A current flows through a telegraph wire from Edinburgh to London, but we do not know whether it comes from Edinburgh or from London. Supposing this knowledge desired, how would you obtain it?

4. A copper wire connects a steel knife and a silver fork, which are both plunged in an orange. The wire is turned so as to lie in the magnetic meridian, the **fork** being to the south and the knife to the north. A small magnetic needle is placed underneath the wire. Will the wire exert any action upon the **needle**? If so, what action?

5. Two compass-needles are arranged near **each** other **so** that both point along the same straight line. A wire connecting **the** platinum and zinc ends of a battery is stretched vertically **half-way** between **the** needles. How will the current **in** the wire affect **the needles,** and how will the result depend upon whether the platinum terminal is connected **with** the upper or lower end **of the** wire respectively?

6. Wires from **two** separate **voltaic batteries** are stretched one above the other from N. **to S.** (magnetic), **and** equal currents pass through both wires. If a magnetic needle, free to turn horizontally but not vertically, is hung half-way between **the wires,** how will it be affected—

   (*a*) **If the currents are** both in the same direction?
   (*b*) **If** the currents are both in opposite directions?

7. **A** number of galvanic cells are connected together in a row so as to form **a battery.** This row **is** laid on a table so **as** to lie N. and S. The zinc is **towards** the N. The poles of the battery **are** connected together **by a wire,** which passes from one pole up one wall of the room, across the **ceiling,** and down the opposite wall to the other pole of **the** battery. How will a magnetic needle be affected **which** is placed under **the** table and just below the battery?

8. A current passing though a long wire is so **weak** that, when the wire is stretched over, and parallel to, a suspended magnetic needle, the needle is not perceptibly deflected. **Describe** and explain an arrangement which would enable you to obtain **a movement** of the needle by the action of the current.

9. **How is it that a galvanometer** with astatic needles is more sensitive **than the same** instrument would be if furnished only with a single needle?

10. Describe and explain an ordinary multiplying galvanometer.

11. **A** steel fork and a **steel** knife are connected by wires with a galvano-**meter.** The knife and fork are used to cut a juicy and well-salted beef-**steak.** What will be the effect on the galvanometer? What will be the

effect when a silver fork is substituted for the steel one, the steel knife being retained ?

Magnetic condition of a wire conveying a current.—Exp. 225.

—Attach a copper wire to the plates terminating a Grove's battery of 5 cells. Bring iron filings in contact with the wire. Observe that they are attracted, and cling in masses round the wire. On breaking contact the filings immediately fall.

We thus learn that a wire, through which a current is passing, becomes a magnet; it has, therefore, a region through which its influence extends.

We will now examine the *magnetic field* due to a current.

Exp. 226. Pass a wire vertically through a hole in a sheet of cardboard (fig. 191). Connect the ends with a 5-cell Grove's battery. Shake iron filings from a muslin bag upon the cardboard. Gently tap it, and observe that the filings arrange themselves in circles round the wire.

The reason of this circular arrangement of the filings will be understood from the following explanation.

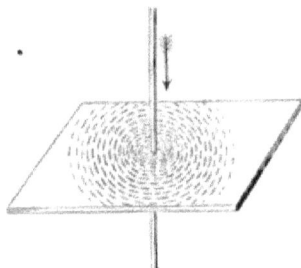

FIG. 191.

Suppose a wire through which a current is passing to enter the plane of the paper at A, fig. 192. Under the inductive influence of the wire the filings become small magnets, and they therefore tend to set themselves at right angles to the wire conveying the current.

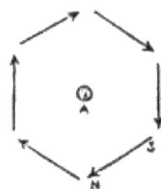

FIG. 192.

According to Ampère's rule, if the observer suppose that he swims in the current, he would pass through the plane of the paper head foremost; if, therefore, he looks at the filing NS (fig. 192), the N-seeking pole will turn in the direction of his left hand, as indicated in the diagram. A similar result will take place with the other filings, so that they lie round the wire.

We thus learn that the direction of the *lines of force of a current* lies in *circles* round the wire.

Magnetisation by current.

—As already indicated on p. 33,

when a current passes through a coil having an iron **bar inserted in** it, the **iron becomes** magnetised. We **must now treat of this** subject more fully.

Spirals or Helices.—Helices are of two kinds : **1,** right-handed ; **2,** left-handed.

Exp. 227. Take a piece of wire or string, and place one end of it on the front of a vertical pencil. Hold it securely in this position with the thumb of the left hand, and with the other hand bend it to the *right* so as to wind upwards. This forms a **right-handed helix** (fig. 193, A).

Again, holding the pencil and the wire or string as before, bend it towards the left and wind upwards. This forms a **left-handed helix** (fig. 193, B).

Observe that these spirals remain the same whichever end is upwards—a right-handed helix cannot be made a left-handed one by merely turning it round. They are like gloves ; one is always right-handed, the other left-handed, unless, indeed, they are turned inside out. Similarly, the only method of converting a right-handed helix into a left-handed one is to turn it inside out. This can be done as follows :—

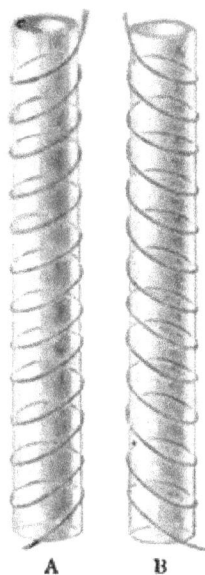

A   B
FIG. 193.

Exp. 228. Make a flat spiral, and show that it becomes a right, or left-handed conical spiral according as its **centre is** pulled out or pushed in.

An ordinary screw (e.g., a corkscrew, fig. 194) is a right-handed spiral, and we can easily imagine the action by which the end of a corkscrew is driven into a cork. This is an excellent method of ascertaining the kind of spiral we are dealing with.

Take the right-handed helix (fig. 193, A), and imagine that you are holding one end ; we can easily see that the other end would be driven into a cork by an action similar to that used with an ordinary corkscrew.

When it becomes necessary to *draw* either kind of spiral we can readily do it as follows :—

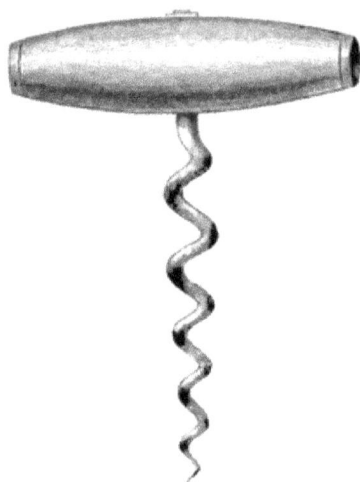

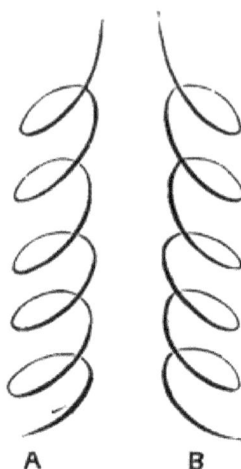

Fig. 194.

Fig. 195.

With a pen or pencil draw a thick curved line from *right to left*, with the concave side downwards, as in fig. 195, A. Continue the drawing as in the diagram, making thick lines, which represent the front of the spiral, from right to left. This forms a *right-handed helix*.

If the thick lines be drawn from *left to right* we obtain a *left-handed helix* (fig. 195, B).

Exp. 229. Attach a right-handed helix, made of copper wire, either insulated or wound on a glass tube, to the wires from a Grove's battery, and place an iron bar within it. Observe, by bringing an unmagnetised iron rod in contact with the ends, that the bar is magnetised.

Before testing the polarity try to discover, by Ampère's rule, the position of the N-seeking pole.

Reason as follows :—

(1) I must know the direction of the current. I ascertain this by looking at the battery and noticing the wire from the free platinum. (2) I follow the direction of the current round the coil. (3) Choosing any particular part of the wire, I

imagine myself swimming in the current so as to look at the bar. The N-seeking pole will lie towards my left hand. (4) I find, then, that in a right-handed helix, the N-seeking pole of the bar is at the end where the current leaves the helix.

Exp. 230. Prove that your reasoning is correct by means of a magnetic needle.

Exp. 231. Repeat the last experiment with a left-handed helix, and prove that the N-seeking pole is at that end of the bar where the current enters the helix.

FIG. 196.

If the direction of the winding be altered occasionally, we obtain a helix represented in fig. 196. By means of a helix of this kind, we may obtain a magnet having consequent poles which are situated at the points where the change of direction takes place.

Exp. 232. Place a steel knitting-needle in such a helix, and after allowing the current to pass for a few minutes, test the position of the consequent poles by means of iron filings. A result will be obtained similar to that in fig. 40.

Exp. 233. Examine an electro-magnet (p. 33) to ascertain the nature of the helix. Then fix the direction which the current is to take, and, reasoning from Ampère's rule, mark the N-seeking pole with gummed paper. Pass a current through the coil, and by means of a magnetic needle test the accuracy of your reasoning.

Exp. 234. Connect an electro-magnet with a voltaic battery. After placing a bar of soft iron across the poles, test the lifting power of the magnet by allowing someone to pull *steadily* at the bar.

Exp. 235. Place a soft iron nail at right angles and close to a wire through which a current from a voltaic battery is passing. By means of iron filings show that the nail becomes magnetised. Deduce the polarity from Ampère's rule, and prove, by means of a magnetic needle, that your reasoning is correct.

Exp. 236. Make a close helix of many turns of guttapercha-covered copper wire. Bend the ends along the helix, and bring them out near the middle. Such an arrangement is called a **solenoid**.

Pass the wires through a large cork, and solder a zinc plate to one end, and a copper plate to the other. Float the apparatus in dilute sulphuric acid. A current passes through the coil in the direction indicated by the arrows (fig. 197).

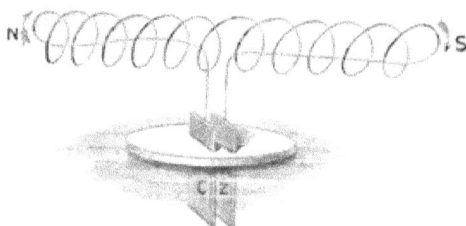

It will be found that the solenoid acts like a magnet, that end becoming S-seeking

FIG. 197

where the current passes in a clockwise direction when the observer looks at that end. Present the S-seeking pole of a magnet to this end. Repulsion ensues.

EXERCISE XXXVI.

1. Suppose the wire that connects the two ends of a voltaic battery be dipped in iron filings, what occurs?

2. Give a clearly-described example of the magnetic action of an electric current.

3. A long copper wire covered with silk is wound several times round an iron rod. On connecting the ends of the wire, one with each terminal of a Daniell's battery, the iron becomes a magnet. How does the magnetisation of the iron (or position of its N-seeking and S-seeking poles) depend upon how the copper wire is wound, and which end of it is connected with the copper end of the battery? Give a drawing.

4. Suppose you wind an insulated wire round a poker and send a voltaic current through the wire, what occurs? What will be the difference between the condition of the knob and the other end of the poker?

5. If you were given any galvanic cell you chose, wire with an insulating covering, and a bar of soft iron, one end of which was marked, state exactly what arrangements you would make in order to magnetise the iron so that the marked end might be a N-seeking pole. Give a diagram.

6. A coil of wire is placed north and south upon a table. An electric current is sent through the coil, which, to a person placed at the south end of the coil, circulates in the direction of the hands of a watch. What is the magnetic condition of a bar of iron placed within the coil?

7. What is meant by an electro-magnet? By aid of a diagram show the polarity of such a magnet.

8. You are required to make a strong electro-magnet. How will you proceed? In what particulars does an electro-magnet differ from an ordinary steel magnet?

9. A piece of copper wire is wound spirally round a ruler from end to end, and the ruler is hung horizontally so that it can turn about its centre while a current is passing through the wire. How can you tell, by using a bar-magnet, in which direction the current is flowing?

10. An insulated copper wire is wrapped round a glass tube, A B, **from** end to end, and a **current is sent through it, which,** to an observer looking at the end A, appears to go round **in** the same direction as the hands of a watch. A rod of soft iron is held **(1)** inside the tube and (2) outside, but parallel to the tube. What will be **the** magnetic pole at that end of the bar which **is** nearest to the observer **in each case?**

11. What **happens** to a fixed **bar of soft iron if the** current from a battery passes above **it and** at right angles to it?

CHAPTER XIX.

RELATION BETWEEN CURRENTS, AND BETWEEN MAGNETS AND CURRENTS.

MAKE two flat spirals (fig. 198), with guttapercha- or silk-covered copper wire (fourteen or fifteen feet of wire will be required for each spiral), in the following manner :—

(1) Take a circular piece of cardboard (about six inches in diameter), and cut a hole (an inch in diameter) at the centre.

(2) Fasten the wire by means of silk thread to the edge of the central hole, leaving ten or twelve inches free, as shown in the diagram.

(3) Wind the remainder so as to form a spiral, fastening each turn securely to the cardboard ; for this purpose it has been found advantageous to use four needles, each one being 90° from the next, and to sew from the centre outwards.

FIG. 198.

(4) Finish the winding at a point on the outside edge so that the two free ends are parallel and close together.

When these spirals are used, e.g. in Experiments 237, 239, 240, 241, *they will be represented diagrammatically, as in figures* 199, 200, &c.

Exp. 237. Attach one end of the coil A by means of connecting-screws to one end of the coil B, so that the current flows in the *same* direction in both coils (fig. 199). Hang the spirals parallel to one another and *very* close together. Connect the free ends with a voltaic battery of five or six cells. Observe that attraction ensues between the spirals.

Exp. 238. Attach the same end of A to the other end of B (fig. 200). The current now flows in *opposite* directions in the coils. Hang the coils parallel to one another and in contact. Notice repulsion.

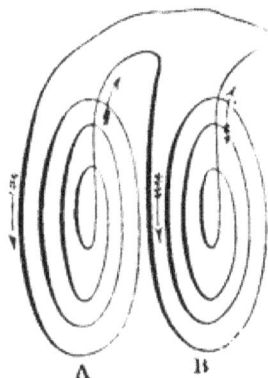

FIG. 199.

FIG. 200.

We therefore learn—(*a*) when two parallel currents flow in the *same* direction through two wires, they *attract* one another ; (*b*) when two parallel currents flow in *opposite* directions through two wires, they *repel* one another.

Roget's Vibrating Spiral.—The attraction of parallel currents flowing in the same direction is well illustrated by means of a spiral (fig. 201). It should be made of 20 or 30 turns of thin copper wire, suspended from a suitable support. The lower end just dips into mercury, which fills a hollow made in the base of the support.

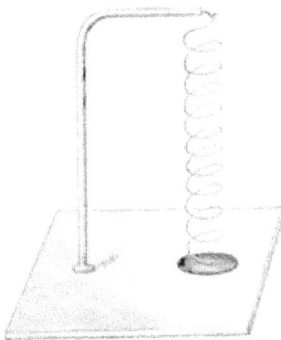

FIG. 201.

Exp. 239. Place one wire from a battery of five cells on the top of the coil, and the other wire in the mercury. A current, therefore, flows through the coil. Observe that attraction takes place between the wires, so that the lower end of the coil is lifted out of the mercury. By this action the circuit is broken and the spiral drops back to its first position. Attraction again takes place, and so on, thus giving an up-and-down motion to the coil.

Induction currents.—Induced currents are instantaneous currents produced in a closed circuit by the influence of currents or of magnets. The currents which produce them are known as *inducing currents*.

Exp. 240. (*a*) Ascertain the *direction* of deflection of a needle in an astatic galvanometer, when a current flows in a particular direction through a spiral (described on p. 202). This may be done by fixing one end of the spiral to a binding-screw of the instrument, and another wire to the other binding-screw; then connect the free ends (of the spiral and wire) respectively with a small piece of copper and zinc, having flannel moistened with dilute sulphuric acid between them.

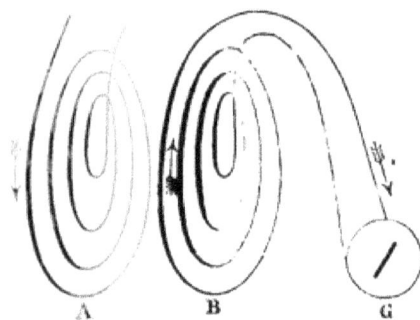

FIG. 202.

(*b*) Placing this spiral (B, fig. 202) on the table, attach its ends to the binding-screws of a galvanometer. Attach the ends of the other spiral, A, to the terminals of a five- or six-cell Grove's battery. Notice the direction of the current round A. Now bring A over B, and observe the direction in which the needle moves, caused by an instantaneous current flashing round B. Knowing the direction of deflection of the needle when the current passes in a particular direction round the spiral, B, we learn that the induced current in B is in the opposite direction to that in A.

(*c*) Lift A from B, the needle moves in the opposite direction, showing that a current flashes round the coil B in the same direction as that in A.

(*d*) Remove one end of A from the battery. Place A upon B and then again complete the circuit. The direction of the needle shows that the current moves round B in a direction opposite to that in A.

(*e*) Keeping A on B, break contact; the current in B then passes round in the same direction as in A.

In these experiments A is called the *primary coil*, and the current in A is called the *primary current*. B is called the *secondary coil*, and the current is called the *secondary current*.

The induced **current** is called *direct* when it moves in the same direction as that in the primary. It is called *inverse* when it moves in a direction opposite to that in the primary.

The experiments on induced **currents** may be admirably performed by means of the apparatus shown in fig. 203. On

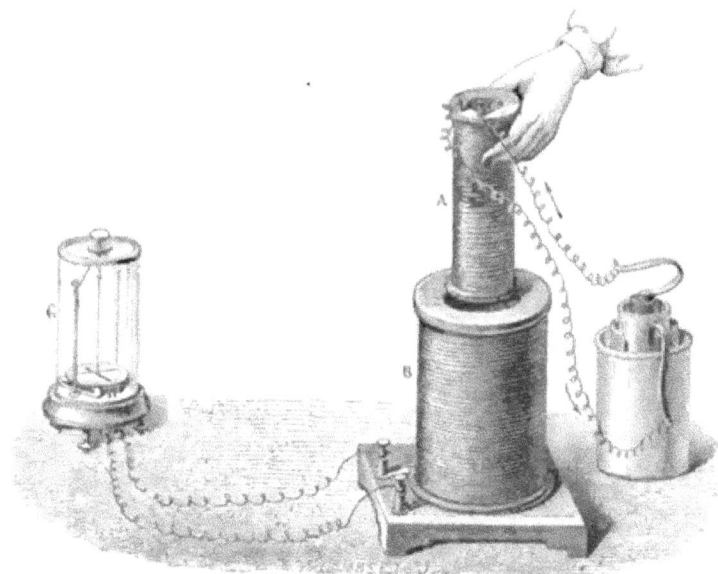

FIG. 203.

a wooden cylinder A, a quantity of silk-covered copper wire is wound (No. 16 B. W. G.). On the large hollow coil, B, a much larger quantity of fine silk-covered copper wire is wound. The primary coil A is connected with the wires from a battery by means of binding-screws, and the secondary coil B is in metallic connection with an astatic galvanometer.

Placing A inside B, the *inverse* secondary current can be obtained by attaching one wire from the battery to one binding-screw on A, and then touching the other binding-screw with the other wire from the battery. The *direct* secondary current is, of course, induced when contact is broken. The diagram represents the method of producing the secondary current by approach or withdrawal of the primary. On placing the coil A inside the coil

B, an *inverse* current flashes round B ; on withdrawing A, a *direct* current is induced.

If the *strength* of the primary current is *increased*, an *inverse* current is induced in the secondary coil ; if *diminished*, a *direct* secondary current is induced. From these experiments we therefore obtain the following results :—

Instantaneous *inverse* currents are induced in the secondary coil by the primary	Instantaneous *direct* curents are induced in the secondary coil by the primary
(*a*) when beginning, (*b*) when approaching, or (*c*) when increasing in strength.	(*a*) when ending, (*b*) when receding, or (*c*) when diminishing in strength.

Extra Current.—Mention must be made of the action in two adjacent wires of a coil when a current flows through them. We can easily see, by reference to fig. 204, that the current in one wire acts on an adjacent wire in the same way as a primary acts on a secondary —e.g., if a wire be coiled as shown in the diagram, and a current moves from P to Z, the part A induces an instantaneous secondary current in the part B, which, moving in the opposite direction, tends to weaken the primary—i.e., on making contact the induced secondary current is inverse. Again, on breaking contact a direct secondary is established, which travelling in the same direction as the broken primary increases its strength. *The direct induced current in the primary wire itself, which strengthens the current when it is broken, is called the* **extra current.**

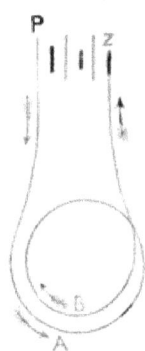

FIG. 204.

Induced currents by magnets.—Exp. 241. Place one of the spirals, B (p. 202) on a table and attach the two ends to a galvanometer, G (fig. 205). Bring the N-seeking pole of a powerful magnet near the coil. A current is induced, which, we learn from the direction of deflection of the needle, flashes round the coil in a direction similar to that in Experiment 240, *b*. The N-seeking pole of a magnet, therefore, behaves like the primary coil A in that experiment, where the current flows in a direction contrary to the hands of a clock (i.e. counter-clockwise). **If** the S-seeking pole be substituted, a secondary **current** passes round in the opposite direction—i.e., the S-seeking **pole** of a magnet has the same **effect as a** coil round which **a current is mov-ing round it in a clockwise** direction.

This experiment can, of course, be performed **with** the apparatus shown in **fig. 203, if** the **primary coil** A **be replaced by a strong magnet.**

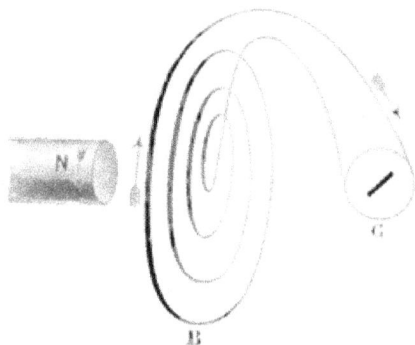

FIG. 205.

Ampère's theory of Magnetism.—From the fact that a solenoid (Exp. 236) acts in every respect like a magnet, Ampère propounded a **theory that** magnetisation is due to current-circulation. This **theory supposes** that every molecule of a magnetic substance **has a current** flowing round it. Before magnetisation (as mentioned **on** p. 9) the currents **move** irregularly ; during magnetisation they assume parallel directions, and the **more per-fect** the magnetisation the more parallel they become. In fig. 18 the currents, moving round the molecules, are represented **by** arrows. The combined **effect** of these molecular currents is equal to **that of a single current** flowing **round** the outside of the magnet. We have shown, in Exp. 241, that this **current** at the **S-seeking pole of a magnet is in the** direction of the hands of a clock (when that pole is looked at), and that at the N-seeking pole is in the contrary direction. The truth **of this theory is also emphasised** by the experi-

FIG. 206.

ments just performed, in **which** currents are induced by the poles of a magnet in a manner similar to those induced by currents. If, therefore, we imagine that currents flow in this manner **round a magnet, which for the sake of simplicity is** represented as a cylindrical one, they will move **as** shown in fig. 206. *Looking*

at the S-seeking pole we see that they flow in a clockwise direction,
and looking at the N-seeking pole they move counter-clockwise.
We also learn that the currents are symmetrical with respect to
the whole magnet, but they appear to flow in contrary directions
when we look at the different ends.

We can easily ascertain, without extra memory work, the
direction of the current at either pole, for, applying Ampère's
rule, if we swim round the bar with our face turned towards it,
the N-seeking pole will always be on our left hand.

De la Rive's floating battery is an extremely useful apparatus
for showing the action of a magnet upon a current. It can be made
and used as follows :—

Exp. 242. Fit a light glass vessel (e.g. a beaker) into a cork,
or into a wooden tray (fig. 207). Fasten strips of copper and zinc

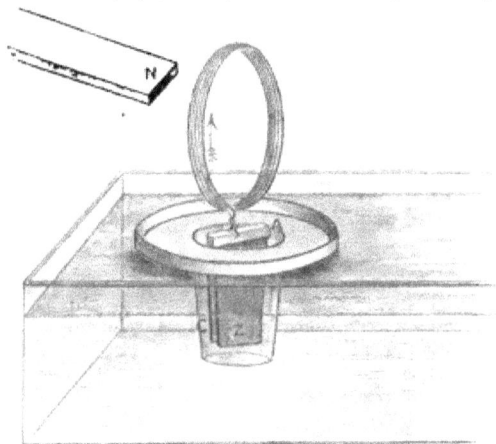

FIG. 207.

(c and z) side by side to a cross-piece of wood, A. Bend silk-
covered copper wire into a coil (say of twenty turns), and pass the
ends through the wooden cross-piece so as to be in contact with
the strips. On filling the beaker with dilute sulphuric acid, a current will pass round the coil. Float the apparatus on water contained in a large vessel. Present the S-seeking pole of a magnet
to that face of the coil in which the currents circulate clockwise.
The battery will be repelled, for the currents in the coil and the
magnet are moving in contrary directions.

If the S-seeking pole be plunged into the coil, the battery will
be repelled, so that it floats off the magnet. It then turns round

and attraction ensues, the coil passing up the magnet until it reaches the middle.

Exp. 243. Present the N-seeking pole to the face of the coil, in which the currents circulate clockwise. Attraction takes place, as in the latter part of Exp. 242. The reason of this effect is due to the fact that the currents in the magnet and in the coil circulate in the same direction.

In Experiment 242 it appears, at first sight, as though the currents in the S-seeking pole flow in a counter-clockwise direction, for we have said that the current in the coil (which is clockwise) and those in the magnet are moving in contrary directions. The reason of this apparent contradiction is, of course, due to the fact that we are not *looking at* the S-seeking pole. The best method of reasoning is this :—

Suppose that the direction of the current is clockwise in that face of the coil which is looked at, and I wish to discover whether attraction or repulsion will take place when a particular pole is presented to the coil.

(1) Consider the S-seeking pole. When I present that pole to the coil, I am looking at the N-seeking pole, and the currents therefore, flow in a counter-clockwise direction. There will then be repulsion.

(2) Consider the N-seeking pole presented to the coil. In this case I am looking at the S-seeking pole, and the currents are, therefore, clockwise. Attraction therefore takes place.

Ruhmkorff's Coil, or **Induction Coil**, is an apparatus by means of which induced currents are produced, having so high an electromotive force that the results given by it are equal to those produced by frictional electricity.

These instruments differ considerably in detail, but their essential parts are similar. The construction will be understood after carefully studying the following diagrams and explanation.

Fig. 208 gives a perspective view of the instrument, while Fig. 209 is a sketch of a dissected coil, having the outer parts removed. Both figures are similarly lettered.

The **reel** consists of a hollow tube of stout paper ; that in the coil from which the diagram is drawn is seven inches long

P

and one inch in diameter. The ends of the reel A A' are made of vulcanite, about $4\frac{1}{2}$ inches square and $\frac{3}{8}$ inch thick. Each

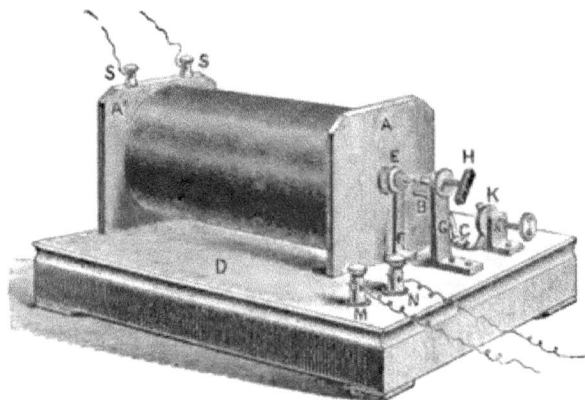

FIG. 208.

end has two holes drilled through it—one in the centre to admit the paper tube, the other through which the primary wire is

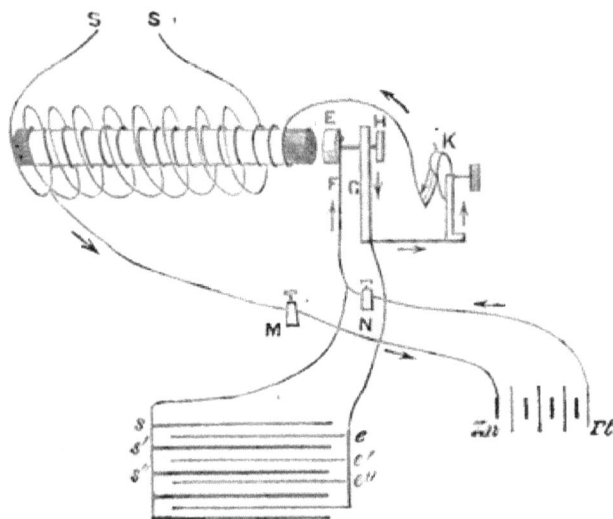

FIG. 209.

passed.   In fig. 208 the primary wire B C is shown passing through the end at the point B.

The primary coil consists of about one pound of No. 16

copper wire covered with cotton. The wire is carefully wound in a close coil round the reel from one end to the other. After several layers are wound, one end of the wire is connected with ĸ, and the other passes through the hole in ʌ′, and is brought back to the binding-screw ᴍ, under the uppermost part of the wooden base ᴅ. The primary is coated with shellac varnish, and is then covered with two or three layers of cartridge paper, which is also varnished. By this means the primary is thoroughly insulated.

The secondary coil generally consists of very fine silk-covered copper wire (No. 36 B.W.G.), about 1,000 yds. in length. It is very carefully wound along the varnished paper which covers the primary, then back again, and so on ; each layer is *thoroughly* insulated by varnishing it and covering it with several sheets of varnished paper. The ends of the secondary are terminated in binding-screws, ss′. Fig. 209 merely gives a *representation* of the secondary by means of the thin lines terminated in ss′. Of course there are many layers of wire.

The coil is then covered with a sheet of velvet or thin vulcanite.

The iron core consists of a bundle of soft iron wires, exactly equal in length. The bundle projects about ¼-inch from the ends of the reel.

The contact-breaker consists of two parts :—

(1) A soft iron head, ᴇ, attached to a spring, ꜰ (figs. 208 and 209.)

(2) A brass upright, ɢ, which carries a screw, ʜ, armed with a platinum point. By means of the screw, ʜ, the distance of the hammer-head, ᴇ, from the end of the iron core can be varied.

The commutator, or current-reverser, is not really necessary,

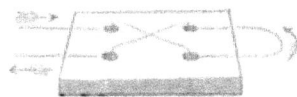

Fig. 210.   Fig. 211.

although it is an extremely useful addition, by means of which the current can be reversed or stopped at any instant.

A very simple form of commutator, from which the principle
of more elaborate ones can easily be understood, consists of a
wooden block containing four small holes which are filled up
with mercury.   The holes are then connected with wires, as
shown in figs. 210, 211.

When these connecting wires are parallel to each other, the
current passes as indicated by the arrows in fig. 210 ; when
the wires cross one another without touching, the current passes
in *a contrary direction* in the *curved* portion of the wire (fig.
211).   If, therefore, a galvanometer be placed in this portion,
the direction of the current through it can be *reversed* at
will.

A more convenient form is represented in figs. 212, 213,
which are similarly lettered.   It consists of a small ebonite or
ivory cylinder, provided with two pieces of brass, E and F,
which gradually narrow off, as shown in fig. 212, so that E is
not in metallic connection with F.   The same pieces are represented in the vertical section, fig. 213.   The cylinder can be

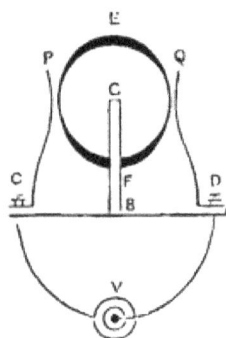
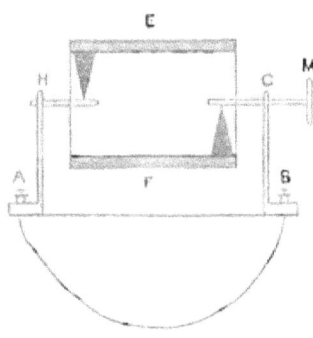

FIG. 212.   FIG. 213.

turned either way on two brass axles, by means of a handle, M,
so that E or F presses on the spring P or Q.

The axles are connected by a metal pin (shown in fig. 213
in a triangular form) with the brass pieces E and F, and are
supported on brass uprights connected with the binding-
screws A and B.   The springs P and Q are connected with the
binding-screws C and D, to which the wires from a voltaic
battery, V, are attached.

Let us now trace the direction of the current when E touches the spring P, and therefore when F touches Q.

The current passes from the battery to the binding-screw C, up the spring P, enters the brass piece E, passes down the pin connected with the axle H, down the upright to A, thence *through the wire to* B, up G, through the axle and pin to F, thence down the spring Q, and so back to the battery.

To reverse the current through the wire joining A and B (fig. 213), all that is necessary is to turn the cylinder so that F touches P, and therefore E touches Q. After the current has passed through P, it enters F, then through the pin and axle down the upright G to B, *through the wire to* A, thence to the axle H, and pin to E, down Q, and so to the battery.

To stop the current the cylinder is placed as shown in fig. 212, so that neither E nor F touches the springs.

In the instrument from which fig. 208 is drawn, the commutator is replaced by a small brass disc, K, part of the edge of which has a projection. When the disc is turned this projection may either come in contact with a spring or not, according as we require the current to be established or broken. The current, however, cannot be reversed by its means.

The condenser is sometimes omitted, but when present it adds very considerably to the effect of the coil. It is placed

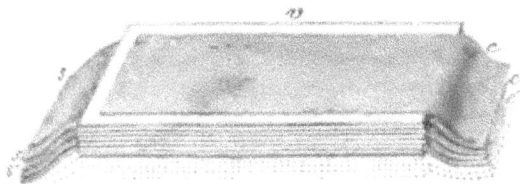

FIG. 214.

in a box forming the base of the coil, and consists of alternate layers of paper, soaked in paraffin (*v*, fig. 214), and tinfoil; thus the condenser consists of sheets of tinfoil separated from each other by a non-conductor, and is, therefore, similar in construction to the condensers described in Frictional Electricity (chap. xiii.)

The sheets of tinfoil project beyond the ends of the paper,

one set at *ss's''*, and the other at the other end *ee'e''*, so that the odd numbers form one coating of a condenser (which is connected with the hammer E ; figs. 208, 209), and the even numbers the other coating (which is connected with the commutator, K). The effect of the condenser is due to the fact that the extra current (see p. 206) passes to a great extent into the condenser, instead of producing the usually brilliant spark between E and the end of H. Upon making contact, this 'bound' electricity escapes and strengthens the current in the primary coil.

Action of the coil.—Following the direction of the arrows, fig. 209, the current enters at the binding-screw N, ascends the spring F, through the screw to G, thence to the commutator K, and so passes through the primary back to the binding-screw M and the battery.

When the current passes through the primary, two important results take place—

(1) a momentary inverse current is induced in the secondary coil ;

(2) the core is magnetised.

When the latter occurs, the hammer-head E is attracted, and the current is therefore broken ; immediately producing a momentary direct current in the secondary.

When the current in the primary ceases, the core loses its magnetism, so that E falls back, which at once re-establishes the current.

These results are repeated during the whole time the current passes.

The induced currents thus obtained give results similar to those produced by frictional electricity—i.e., they are capable of yielding sparks by overcoming the resistance of the air between the terminals. For this reason the instrument is sometimes called an **Intensity Coil**. We may, therefore, perform experiments with a Ruhmkorff's coil similar to those performed with frictional electricity. The effects may be divided into physiological, mechanical, heating, luminous, and chemical.

Physiological effects.—The shocks received from a Ruhmkorff's coil are exceedingly dangerous, unless a very small coil is used.

Mechanical effects.— Exp. 244. Hold a piece of paper or cardboard between the terminals of the secondary coil. Notice that the paper is perforated, a slight burr appearing on both sides in a manner similar to that in Exp. 185.

Heating effects.— Exp. 245. Attach a very fine iron wire to the terminals. When contact is made, observe that the wire fuses.

Luminous effects.— Exp. 246. Place the terminals of the secondary near together. Notice that a series of sharp, bright sparks are produced. The length of spark varies with the size and power of the instrument. Mr. Spottiswoode's Induction Coil (the largest yet made) gave a spark 42 inches long, when worked with 30 Grove's cells. It had a secondary coil 280 miles long.

Exceedingly beautiful effects are obtained when the discharge from a coil is sent through rarefied gases or vapours.

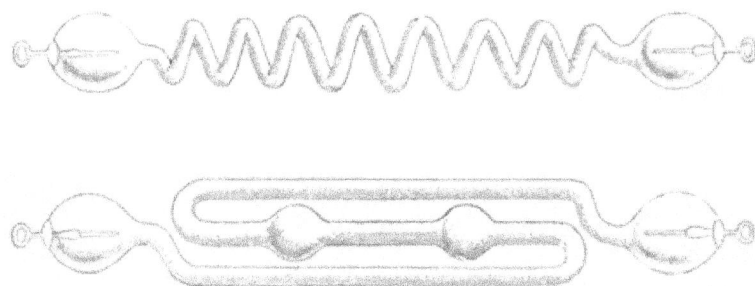

FIG. 215.

For this purpose **Geissler's tubes** are used. These tubes are made of glass of various shapes, and are filled with different gases or vapours, which are then exhausted so that the pressure is extremely small. A platinum wire is fused through each end, the outer portion of which is made into a small loop, while the inner passes a short distance into the tube. Two of these tubes are represented in fig. 215.

Exp. 247. Connect the two platinum wires with the terminals of the secondary coil. On making contact observe the luminous striation.

The striæ vary in lustre, form, and colour, according to the degree of exhaustion, the kind of gas or vapour, and the size and shape of the tube.

FIG. 216.

Fig. 216 shows the striæ when carbonic acid is exhausted so that the pressure is only $\frac{1}{30000}$ of an atmosphere. The colour is greenish.

Chemical effects.—Exp. 248. First fill a eudiometer with mercury, and then allow either the detonating gas—the two gases, oxygen and hydrogen—given off in the electrolysis of water, or, as in Exp. 183, two volumes of hydrogen and one of oxygen, to enter the tube. Attach the loops of the eudiometer, by means of wires, to the terminals of the secondary coil. On passing the spark, observe the film of moisture on the sides of the tube due to the formation of water. On opening the bottom, the tube is filled with mercury.

EXERCISE XXXVII.

1. You have a metal hoop. Describe, by aid of a diagram, some arrangement by which, without touching the hoop, you could make electric currents pass round it, first one way and then the other.

2. What do you know respecting the laws of attraction and repulsion between currents? What apparatus should you use to prove them?

3. A piece of covered wire is passed a few times round a wooden hoop; its ends are joined up to a galvanometer. The ends of another piece of covered wire which is wrapped round a similar hoop are joined up to a

battery. What will happen if the two hoops are (1) brought quickly near to one another, and (2) if they are quickly separated?

4. I suspend a magnet by means of a silk thread tied round its centre. I attach the ends of a coil of wire to a galvanometer. I then approach the coil with the magnet. What happens, and why?

5. Describe a Ruhmkorff's coil.

6. What are *Geissler's tubes*? Describe an experiment you have performed with them.

CHAPTER XX.

ELECTROLYSIS.

Exp. 249. Fit up a Grove's battery of 4 or 5 cells. Attach one wire (say that from the platinum end) to a binding-screw of a galvanoscope. Place the end of the wire from the zinc plate in a small vessel containing turpentine or petroleum. Complete the circuit by means of a wire passing from the liquid to the other binding-screw of the galvanoscope. Observe that the needle is not deflected unless the two wires in the liquid are in contact. We, therefore, learn that some liquids are non-conductors.

Exp. 250. Repeat the last experiment with mercury instead of turpentine or petroleum. The needle is now deflected, proving that the mercury is a conductor. On examining the mercury we find that it is unaltered. It is, in fact, an element, and as such is incapable of decomposition.

Exp. 251. Dissolve a few crystals of potassium iodide in water contained in a test-tube. As in Experiment 249, place the wires without touching in the solution. Observe (1), by means of the galvanoscope, that a current is flowing ; (2) that bubbles of gas begin to rise from the wire in connection with the zinc ; and (3) that the liquid becomes brown. This brown coloration appears most round the wire from the platinum, and is due to the liberation of the iodine. The salt is *decomposed* into potassium and iodine, as shown by the following equation

$$KI = K + I.$$

The liberated potassium, however, immediately acts upon the water, forming caustic potash and liberating hydrogen

$$K + H_2O = KHO + H.$$

It may be mentioned, in passing, that an excellent test for weak currents depends upon this fact. It will be understood from the following experiment.

Exp. 252. Make a solution of potassium iodide in water. Soak a piece of white blotting-paper with the solution. Place the wires from a single cell on the paper an inch or two apart. Brown marks will appear wherever the wire from the platinum end touches the paper.[1]

If the solution has a small quantity of starch dissolved in it, the paper will be turned to an indigo blue colour at the point of contact. This coloration is due to the action of the iodine on the starch.

From Experiments 250, 251 we learn that bodies which conduct electricity may be divided into two classes—

(1) those incapable of decomposition.

(2) those capable of decomposition.

Explanation of terms.—The decomposition of conducting compounds by the electric current is called **electrolysis**. The compound itself is known as the **electrolyte**. The ends of the wires from the battery which act upon the electrolyte are termed **electrodes** (A and B, fig. 217). The positive electrode—i.e. the pole where

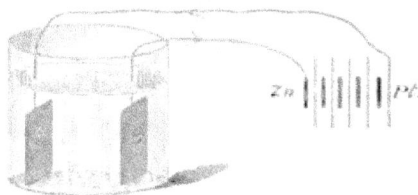

FIG. 217.

the positive current enters the electrolyte—is called the **platinode** or the **anode** ; the negative electrode, the **zincode** or the **cathode**. In fig. 217, A is the anode, B is the cathode. The substances given off at the electrodes are called **ions** ; the one appearing at the anode is called the **anion**, that at the cathode is called the **cation**.

Electrolysis of Water.—To electrolyse water we require an apparatus similar to that represented in fig. 218. It consists of a glass vessel fixed on a wooden base. Two platinum electrodes, *h n*, are fitted to the bottom of the vessel, and are connected with copper wires terminating in binding-screws.

---

[1] A result similar to this was obtained with frictional electricity (p. 146). Faraday used this test to establish the identity between voltaic and frictional electricities.

This part of the apparatus can be satisfactorily made as follows :—

Obtain a glass funnel, which measures five or six inches across the top.   File off the stem at a point about half an inch

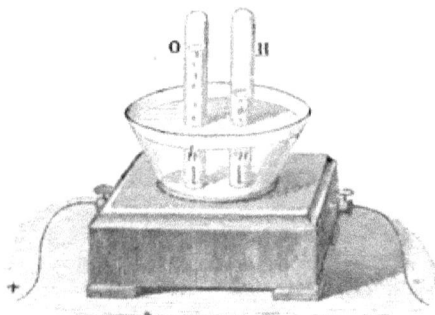

Fig. 218.

from the bottom of the funnel.   Solder strips of platinum foil to two copper wires, and then pass the latter from the inside of the funnel through the stem. Arrange the strips parallel to each other, and then fill the stem and part of the funnel with plaster of Paris, so that the pieces of platinum project above it.   Take care that the wires are not in contact.   Now melt some paraffin and pour it over the surface of the plaster of Paris, so as to make the apparatus water-tight.   The vessel must now be placed on a suitable support—e.g., a retort-stand.   Two test-tubes of equal size are also required.

Exp. 253. *To use the apparatus.*—Fill the vessel three parts full of water to which a little sulphuric acid has been added, in order to increase the conductivity (pure water is a bad conductor of voltaic electricity).   Fill the two test-tubes with acidulated water, and invert them over the electrodes.   The free ends of the wires are connected with a Grove's cell (which is sufficient for this purpose), or preferably with a battery of four or five cells, with which the action proceeds rapidly.   Observe that bubbles of gas immediately begin to rise from the electrodes.   Allow the action to go on for a short time, and notice that the volume of gas in the tube connected with the cathode (H, fig. 218) is nearly double that in the tube O, connected with the anode.   When the tube H is full, test its contents by bringing a lighted match near the mouth of the tube. Notice that the gas burns with a pale blue flame—this is one of the tests for hydrogen.   When the tube O is full, plunge a glowing splint of wood into it, observe that the wood bursts into flame—a test for oxygen.

Theoretically the volume of hydrogen should be exactly double that of oxygen, as shown in the following equation

$$H_2O = H_2 + O.$$

Practically, however, the volume of oxygen is slightly less than that obtained from this consideration ; the chief reasons of this difference arise from the facts (1) that oxygen is more soluble in water than hydrogen, and (2) a portion of the oxygen is evolved in a denser form as ozone.

An apparatus of this kind which *measures* the volumes of the gases is called a **Voltameter**.

Exp. 254. Collect the hydrogen and oxygen together in a strong glass bottle—e.g., a soda-water bottle—or transfer them from the tubes used in the previous experiment into the bottle. (This transference must be done by filling the bottle with water, and bringing the tube full of gas to the mouth of the bottle, which is of course kept below the surface of the water. On gradually inclining the tube, the water is displaced by the gas.)
Bring a lighted taper near the mouth ; a violent explosion occurs, due to the recombination of the two gases to form water. The water can be seen as a film of moisture on the interior surface of the glass. It is advisable to wrap the bottle, even if it is a strong one, in a duster, as the explosion is so violent that the bottle may burst, and so perhaps cause a serious accident.

Electrolysis of Hydrochloric Acid.— Hydrochloric acid may be electrolysed by means of a similar apparatus. The electrodes, however, must be made of carbon, as platinum is attacked by chlorine in a nascent [1] state. An excellent form of apparatus for this purpose—and indeed for water—is that shown in fig. 219. It is known as *Hofmann's Volta-*

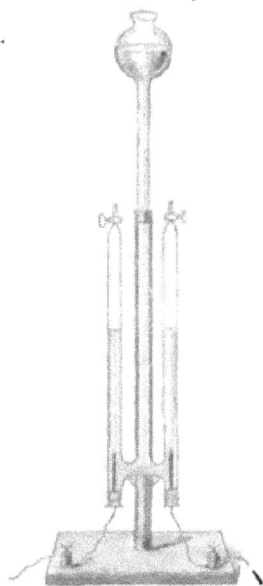

Fig. 219.

[1] The term *nascent* in chemistry refers to the state of a body at the instant of being evolved from a compound containing it. At such an instant it has more active properties than when it exists in a free state.

meter. It consists of two glass tubes (often graduated), bent into a U shape. At the bottom of each tube is a pointed carbon rod in connection with a platinum wire, by means of which connection is made with the battery. The other ends of the tubes are open, but can be closed at pleasure by stop-cocks. The U-tube is connected at the bottom with a straight tube terminated by a funnel. Into this tube the hydrochloric acid is poured ; on opening the stop-cocks the liquid fills the apparatus.

Exp. 255. Fill the voltameter with the acid, and then close the stop-cocks. As chlorine is extremely soluble in water, add a quantity of common salt, which has the effect of diminishing this solubility. Allow the action to continue for some time, and observe that we obtain nearly equal volumes of gas in each tube. Theoretically the volumes are exactly equal, according to the equation—

$$HCl = H + Cl.$$

Observe that the gas given off at the anode is of a yellow colour, and on opening the stop-cock it has a peculiar odour, which irritates the air-passages and lungs. These are characteristics of chlorine. Bring a lighted match to the gas given off at the cathode. It burns with a blue flame—a test for hydrogen.

The electrolysis of all *binary compounds*—i.e., compounds consisting of two elements—is similar to that of water and hydrochloric acid ; one element is given off at the positive electrode, and the other at the negative electrode.

Electro-Chemical Series.—Those elements which are given off at the positive electrode are called electro-negative elements, those from the negative electrode, electro-positive elements.

In the following list, any two elements being taken, the one standing last is electro-positive to the other.

Oxygen	Hydrogen	Tin
Sulphur	Gold	Lead
Nitrogen	Platinum	Iron
Chlorine	Mercury	Zinc
Iodine	Silver	Sodium
Carbon	Copper	Potassium

It is important to remember that, in all our examples on electrolysis, *hydrogen and metals are evolved from the negative electrode.*

Electrolysis of salts.¹—(A) Copper sulphate. Exp. 256.
Fill a U-tube, fig. 220, with a solution obtained by dissolving
crystals of copper sulphate in water. Pass platinum plates connected with the wires from a cell or battery into the arms of the
tube. Observe that copper is deposited on the negative electrode,
and bubbles of oxygen gas are set free at the positive electrode.

What actually happens in the tube is this :—

When the copper is set free from the salt, the 'sulphion,'
SO_4, is liberated, which, being incapable of existing alone,
at once attacks the water, taking hydrogen from it (forming
H_2SO_4), and setting the oxygen free.

These reactions are expressed by the following chemical
equations :—

$$CuSO_4 = Cu + SO_4$$
$$SO_4 + H_2O = H_2SO_4 + O.$$

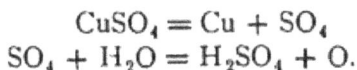

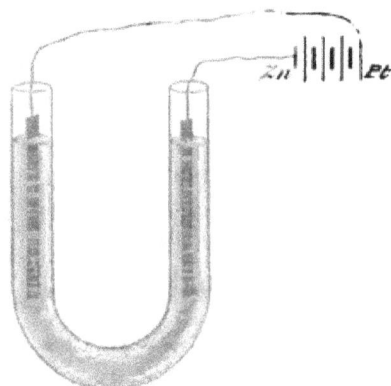

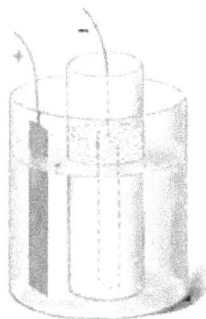

FIG. 220.   FIG. 221.

(B) **Sodium sulphate.—Exp. 257.** Make a strong solution of
sodium sulphate (Na_2SO_4) in water. Place a porous cell inside a
glazed earthenware or glass vessel (fig. 221). Partially fill both
vessels with the solution, and add to each a little litmus solution.
In the porous cell add a drop or two of acid, and in the earthenware cell a drop or two of an alkali (e.g., ammonia).

¹ All *acids* contain hydrogen—e.g., H_2SO_4 (sulphuric acid), HCl (hydrochloric acid). When the hydrogen of an acid is replaced by a metal, a
salt is formed—e.g., $CuSO_4$ (copper sulphate), NaCl (sodium chloride or
common salt).

The experiment depends upon the fact that

    acids turn blue litmus red

  alkalies ,, red ,, blue.

The solution in the porous cell therefore becomes red, that in the earthenware cell, blue.

Plunge the negative electrode into the porous cell (i.e. the red solution) and the positive electrode into the blue solution. Observe that bubbles of gas rise from both electrodes.

It appears that the sodium sulphate is decomposed into sodium and 'sulphion,' thus

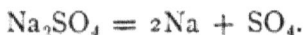

$$Na_2SO_4 = 2Na + SO_4.$$

The sodium (Na) liberated at the negative electrode *immediately* combines with the water, forming caustic soda (NaHO) and setting the hydrogen free. This action is represented by

$$Na + H_2O = NaHO + H.$$

At the same time the 'sulphion' attacks the water, forming sulphuric acid and liberating the oxygen.

$$SO_4 + H_2O = H_2SO_4 + O.$$

The gases which were observed to rise were (1) hydrogen from the cathode and (2) oxygen from the anode.

Again, caustic soda produced at the negative electrode is an alkali, and sulphuric acid produced at the positive electrode is an acid; we ought therefore to find that the red solution has turned blue by the action of the caustic soda, and the blue solution has turned red by the action of the sulphuric acid. On examining the cells we find that this has actually taken place.

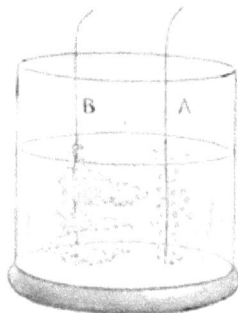

Fig. 222.

(C) **Lead acetate.—Exp. 258.** Partially fill a glass vessel with a solution of lead acetate (sugar of lead). Dip two platinum wires, attached to a two-cell Grove's battery, into the solution, and observe that bubbles of gas (oxygen) rise from the anode, A (fig. 222), while beautiful fern-like branches of lead begin to grow on the cathode, B. In time the space between the electrodes is bridged over and the action ceases. The anode is sometimes coated with a dark-brown powder—lead peroxide, PbO_2—due to the chemical combination of lead and oxygen.

(D) **Silver Nitrate.**—If a solution of silver nitrate be employed as in the last experiment, the silver is deposited on the cathode in delicate filaments.

Discovery of Potassium and Sodium.—By means of electrolysis several substances which were regarded as elements have been shown to be compounds. Davy, by using 250 zinc-copper cells, proved that caustic potash and soda were compounds. His experiment may be repeated with a five-cell Grove's battery,

by making a small cavity in a piece of solid caustic potash, and placing a drop of mercury in it (fig. 223). The caustic soda is then placed on a platinum plate. The terminal from the platinum end of

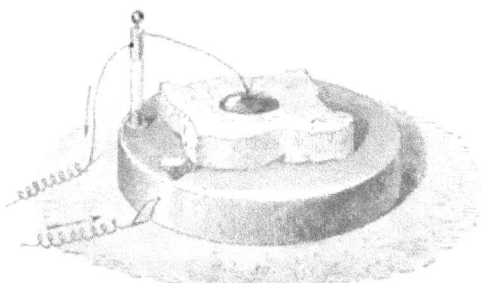

FIG. 223.

the battery is brought in contact with this plate, and that from the zinc end is dipped in the mercury. The potassium is liberated at the negative electrode, and forms an amalgam with the mercury. The mercury is then distilled off out of contact with air. Oxygen is liberated at the positive electrode. The reaction is expressed thus

$$2KHO = 2K + H_2O + O.$$

Electrotyping is an extremely useful process by which we obtain impressions of coins, wood engravings, &c.

The process is as follows :—A *mould* of the object to be produced is first obtained in gutta-percha or plaster of Paris. With gutta-percha or plaster of Paris, it is very important that the face and edges of the cast should be covered very carefully and thoroughly with graphite, so as to make a conducting surface. The mould is then attached to the wire connected with

Q

the zinc end of a Daniell's cell—i.e. it is made the cathode, and a copper plate forms the anode. The electrodes are then immersed in a strong solution of copper sulphate.

As we found in Exp. 256, a layer of copper will be deposited on the mould, the thickness of which, of course, depends upon the time during which the action proceeds.

The copper plate, forming the anode, gradually dissolves as the copper is removed from the solution, thus keeping it at a constant strength.

The reason of taking electrotypes of wood engravings arises from the fact that the carved block would quickly wear away by constant use. After the mould has been taken and a copper surface deposited, the back is filled up with molten type-metal, and when the mould is removed many thousand copies can be printed from the copper impression, while the original block may again be used for the same purpose.

Fig. 224 represents a suitable arrangement for this purpose. The trough is made of glass, slate, or wood, lined with india-

FIG. 224.

rubber or coated with marine glue.[1] Across the trough, which contains an acid solution of copper sulphate, are placed copper rods, B and D, which are respectively connected with the negative and positive poles of a battery.

The mould, *m*, and the copper sheet, C, are then fastened

---

[1] Marine glue consists of caoutchouc dissolved in naphtha. After allowing this solution to stand for about ten days, shellac is added.

to the rods and immersed in the liquid, the distance between them being about two inches.

Electro-plating.—The object of this process is to cover the surface of a base metal—e.g., German silver or copper—with a superior metal—e.g., silver or gold—either for the purpose of protecting them from oxidation (rusting) or to give them the appearance of being wholly composed of the superior metal. Steel bicycles and tricycles, for example, are nickel-plated for both these purposes; the nickel prevents the steel from rusting, and it improves the appearance of the machine.

It is absolutely necessary that the surface of the body to be plated should be a conductor, and it must be thoroughly clean.

The principle will be understood from the following :—

Let us suppose that our object is to coat a German-silver spoon with silver.

The spoon must first be thoroughly cleansed by (1) boiling it in a weak solution of caustic soda to remove any adhering grease, (2) washing with water, (3) immersing it for a moment in dilute nitric acid, (4) brushing it with a hard brush, and (5) plunging it into clean water. The solution, contained in a glass vessel, is made of one part by weight of silver cyanide, one part of potassium cyanide, and 125 parts of water, together with a few drops of carbon bisulphide. Two metal rods are placed over the vessel, from one of which the spoon is hung by means of a wire (waxed all over), and from the other a silver plate is suspended by a silver wire. The rod holding the spoon is then connected with the *negative* pole, and the other with the positive pole of a voltaic cell.

As the silver is deposited on the spoon (which must be turned occasionally), an equal amount of silver dissolves from the plate, forming the positive electrode, thus keeping the solution at a constant strength.

Uses of Electrolysis.

We therefore learn that electrolysis may be used—

(1) to ascertain the chemical constituents of compounds;

(2) to obtain perfectly pure metals (as in Experiments 256 258).

(3) to electrotype ;

(4) to electroplate ;

and it may be also used (5) to ascertain the relative strength of currents. The student will easily understand that if one current, A, acts on water for a certain time, a certain quantity of gas is evolved, which may be measured in a graduated tube : if a second current, B, acting for the same time, causes twice this quantity to be evolved, then the current B is twice the strength of the current A.

EXERCISE XXXVIII.

1. Tell me what you understand by chemical combination and chemical decomposition, illustrating your answers by reference to the formation and decomposition of water.

2. Give me a good example of chemical combination and decomposition brought about by the same current.

3. Describe and illustrate by a sketch how water may be decomposed by a current of electricity.

4. State what you know regarding the substances oxygen and hydrogen ? How would you mix them and deal with them so as to form water ?

5. The current from a voltaic battery is passed at the same time through a thin wire and through dilute sulphuric acid, connected in series. What will happen to the wire and to the dilute acid, and what change (if any) will be produced in each case by reversing the battery connections, so as to alter the direction of the current through the wire and liquid ?

6. A vessel containing a solution of common salt (NaCl), coloured with a little litmus or indigo, is divided into two parts by a partition formed by stitching together several layers of blotting-paper. Two wires coming from the two ends of a Grove's battery are dipped into the liquid on opposite sides of this partition. On one side the colour is observed to disappear. Explain its disappearance, and mention the end of the battery (whether the zinc end or the platinum end) from which the wire that destroys the colour proceeds.

NOTE : *Chlorine has the important property of bleaching vegetable colouring matters in the presence of water.*

7. One end of a copper wire is fastened to the copper end of a battery, and one end of another copper wire is fastened to the zinc end of a battery. What happens if the other ends of these two wires are put side by side (but not touching) into a solution of copper sulphate ?

8. State what you know about the process of electroplating.

9. Draw and describe an arrangement by means of which a penny piece can be coated with silver.

10. How could you obtain an electrotype of a medal?

11. Explain as fully as you can the uses of electrolysis.

12. What is meant by *electrode, zincode, platinode, anode, cathode,* **ion,** *anion, cation.*

13. Give chemical equations to represent the actions which take place when a solution of sodium sulphate is electrolysed.

14. Explain what occurs when the wires from a cell are placed on blotting-paper moistened with a solution of (*a*) potassium iodide, (*b*) potassium iodide and starch.

CHAPTER XXI.

THE ELECTRIC TELEGRAPH.

Use.—The electric telegraph is used to transmit messages by means of a recognised system of signals, from one place to another at a considerable distance apart.

Requisites.—To transmit the message it is necessary to have

(1) *batteries* to generate the currents ;

(2) *circuits*, or lines of wires connecting the two places ;

(3) *instruments* for sending and receiving the signals.   In single needle instruments, these consist of

(a) *a commutator*, for sending the signals from one place ;

(b) *a dial*, for receiving them at the other place.

Batteries.—Daniell's battery is the one generally used in England, a common form of which is shown in fig. 225. It

FIG. 225.

consists of a teak trough (2 ft. long, 6 inches wide, and 5½ inches deep), which is divided into ten cells, separated from each other by slate partitions and coated throughout with marine glue. Each cell is divided by a porous partition into two parts.   A zinc plate, z (fig. 226) is connected by means of a copper strip with a copper plate, which passes over the top of the slate partition. The part of the cell containing the zinc is nearly filled up with water, and that containing the copper is partly filled with crystals of copper sulphate, upon which water is then poured to the same height. Pure water only is added in practice, because zinc sulphate is formed spontaneously in the cell if it be allowed to stand for about two days before use.

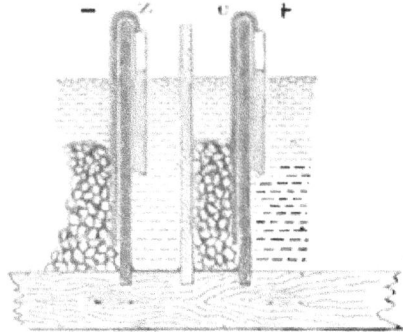

FIG. 226.

Circuits. — Some years ago two lines were employed, one being used for the current to pass to the receiving station, and the other for the current to return to the battery. One wire only is used at the present time, each end of which, as shown in fig. 234, is in metallic connection with a plate sunk to some depth in the ground. The earth acts in a manner similar to the old return wire. It is unnecessary to consider that the same current returns to the battery, the earth merely dissipating the charge owing to the fact that it is at a lower potential than the place from whence the current starts.

Overhead wires are made of galvanised iron to prevent them from rusting, and are insulated by means of porcelain insulators, which are attached to poles or buildings.

Underground wires in towns are made of guttapercha-covered wires placed in iron pipes.

Submarine cables in which great strength is necessary vary somewhat in construction, although the principle involved in each is the same, and will be understood by reference to figs. 227, 228, which are representations of an Atlantic cable. Fig. 228 is a transverse section of the longitudinal piece shown in fig. 227. It consists of a *core* of seven copper wires, insulated by being covered with layers of gutta-percha. This is then covered with hemp, which in turn has a coiled sheath of steel wire placed round it.

At the station from which the message is sent the line is connected with the positive pole of the battery ; the current

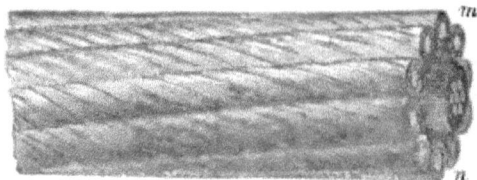

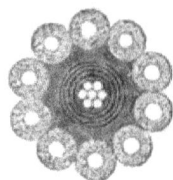

FIG. 227.   FIG. 228.

then passes to the receiving station, where it enters the earth by means of a wire and plate, fig. 234.   This applies to ' single current ' working only.

The single needle instrument.—This consists of a gal-

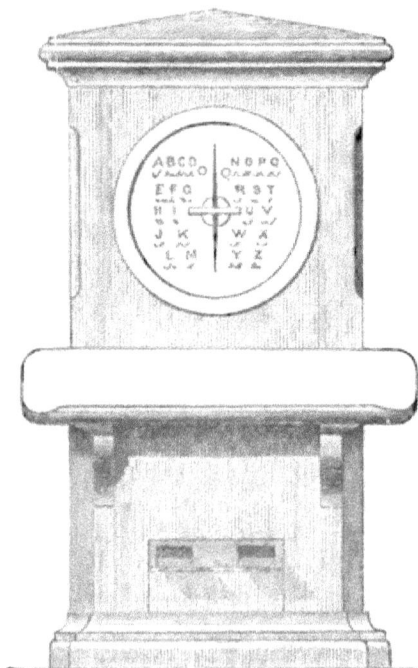

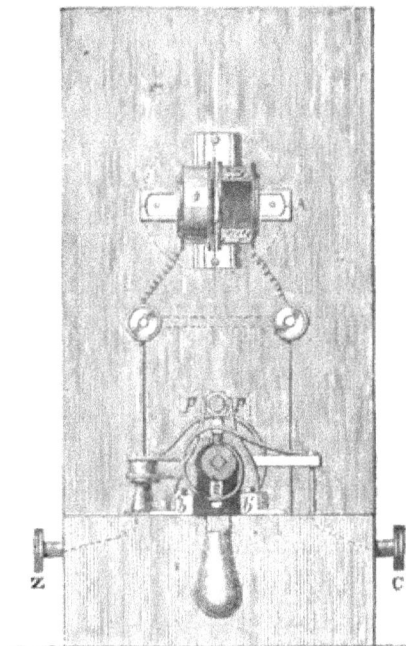

FIG. 229.   FIG. 230.

vanometer fixed vertically in a case having a dial like a clock, and provided with a *commutator*.   Two forms of the instrument are in common use ; the principles, however, of each are the same, the difference consisting in the arrangement of the sending portion, one having a *drop-handle* and the other a *pedal* or *tapper*. Fig. 229 represents the pedal, and fig. 230 the drop-handle form, having the outer case removed.

(*a*) **The galvanometer**, A (fig. 230), consists essentially of a *coil* of fine silk-covered copper wire, in which is placed a vertical *magnetic needle* capable of deflection to the right or left according as a current is sent in one direction or the other round the coil. The magnetic needle is connected with a steel *indicator* placed in front of the dial (fig. 229). One terminal of the coil is in connection with the line, and the other in connection with the commutator to 'earth.'

(*b*) **The Dial.**—The two small circles on the face of the dial, fig. 229 ' represent two *stops* (about one inch apart), which are placed in these positions to prevent the indicator, and the needle with which it is connected, being deflected too widely. The face of the dial has usually the *code of signals* printed upon it—e.g., *t* is indicated by a deflection to the right ; *e* by one to the left ; *m* by two deflections to the right ; *a* by one to the left followed quickly by another to the right, and so on, the letters occurring most frequently being the ones most easily signalled.

(*c*) **The** commutator or *current-reverser* (the principle of which is described on p. 212), as previously mentioned, has two forms :—

(1) The *drop-handle*, which, when hanging vertically, breaks connection with the battery, but when moved to the right or left deflects the needle in those directions.

(2) The *tapper* or *pedal*, which will be easily understood by reference to fig. 231. The tappers E and L are two metallic springs, E being in connection with the earth, and L in connection with the line. They lie between two strips of metal, z c, z being in connection with the zinc end of the battery and c with the copper end.

When the tappers are at rest they press on z. When L is pressed down so that it touches c the circuit is completed, and a current passes from c along the line and coils of the receiving instrument at the distant station, enters the earth, and so returns back to z, which is in connection with the negative pole of the battery. When E is depressed and L is in contact with z, the current passes in the opposite direction, so that the needle at the distant station is deflected in the opposite direction.

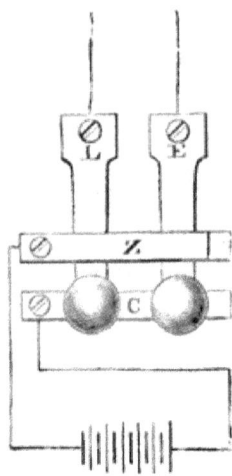

Electromagnetic Instruments. — In these instruments an electro-magnet is employed instead of a galvanometer, as described in the needle instrument. **The Morse instrument** is the one most used at the present time. It consists of (1) the *receiver*, and (2) the *transmitter*.

The Receiver.—The essential part of the instrument consists of an electro-magnet, which, when a current passes round the coil, attracts an armature, *a*, fig. 232, connected with a lever moving about c and between *d* and *e*. It may be arranged as

FIG. 231.

a *sounder*, in which case the person who is receiving the message notices the duration of the clicks produced at *d* and *e*. It may also take the form of an *embosser*, to print the signals upon a strip of paper moved by clockwork through the instru-

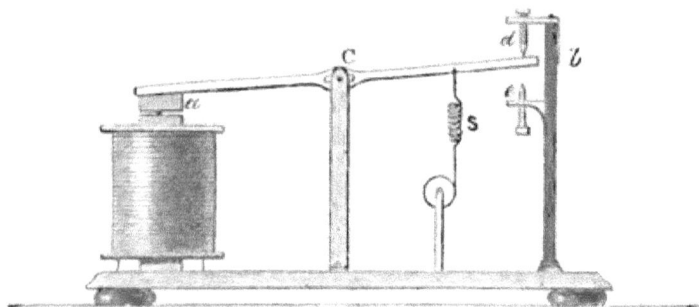

FIG. 232.

ment. The most modern form is that of an *ink-writer*, in which the electro-magnet dips a small wheel into a reservoir of printer's ink and then presses it against a paper ribbon, which is moved as before by clockwork. If the duration of the current is short it prints a dot, if longer a dash,

The following table gives the *dot and dash*, and the **single needle** alphabets.

PRINTING.		SINGLE NEEDLE.	PRINTING.		SINGLE NEEDLE.
A	._	./	N	_.	/.
B	_...	/...	O	___	///
C	_._.	/./.	P	._._.	.//.
D	_..	/..	Q	__._	//./
E	.	.	R	._.	./.
F	.._.	../.	S	...	...
G	__.	//.	T	_	/
H	....	....	U	.._	../
I	..	..	V	..._	.../
J	.___	.///	W	._ _	.//
K	_._	/./	X	_.._	/../
L	._..	./..	Y	_.__	/.//
M	__	//	Z	__..	//..

The Transmitter, or Key, consists of a brass lever *a b* (fig. 233) movable about a horizontal axis. On pressing the knob *n* the lever comes in contact with the button *x*, connected with

FIG. 233.

the positive pole of the battery, so that a current passes through the wire P, *x*, the lever, *m*, and the line wire L, to the distant station, the duration of which depends upon the time the lever is in contact with *x*, thus producing either a dot or a dash. When the pressure is removed a spring lifts the lever, until it touches a button immediately below *b* and in connection with the wire A (leading either to the indicator or to a relay) ; a current, therefore, passes from the line wire L, through *m*, the lever, and the button to the *indicator* or *relay*.

The Relay.—When a current has passed over a long distance its strength has fallen off considerably ; in fact, it may have diminished so much that it fails to actuate the armature *a* (fig. 232) ; it therefore becomes necessary to introduce some contrivance which will give assistance by means of a local current. For this purpose a *relay* is introduced. This consists of a delicate electro-magnet, having a coil of very fine wire. When the weakened current passes round the coil, the electro-magnet attracts an armature, which closes a local circuit, in which a battery and receiver are included. Thus the weak line current brings into action a strong local current, which performs the necessary work.

Arrangement for the transmission of a message.—At one

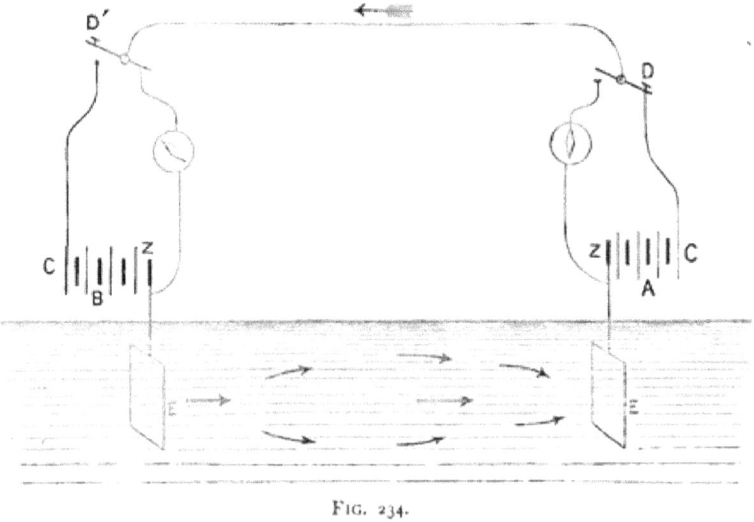

FIG. 234.

station there is a battery, A, fig. 234, the negative pole of which is connected with the earth, and the positive pole with a *key*,

D. When D is depressed (as in the figure) it breaks contact with the receiving instrument at that end, and the current passes along the line, through the key D' and the receiving instrument, where it produces a signal, and thence to the plate E, which is buried in the earth.

The diagram represents the galvanometer arrangement; the principle is, however, the same if a Morse instrument be used.

Electric Bell. — This instrument consists of an electro-magnet, fig. 235, one end of the coil of which is in connection with the binding-screw m, to which is attached the wire passing from the '*push*.' The other end of the coil is connected with a spring, c, attached to an armature, a; this again is pressed by the spring c, connected with the binding-screw n, which leads a wire to the zinc end of a battery of one or two Leclanché cells. The carbon end of the battery is connected by means of a wire with the 'push.'

When the 'push' is pressed, the circuit is completed, and a current flows round the electro-magnet, causing the armature a to be attracted. When this takes place, the clapper, P, being carried by the armature, a, strikes a gong,

FIG. 235.

T; and contact being broken between a and the spring c, the current ceases and the magnet no longer acts. The spring c now comes into play, bringing the armature back to c, thus completing the circuit again. E again attracts a, and so on during the time that the 'push' is pressed.

EXERCISE XXXIX.

1. Give a diagram to illustrate the arrangement by which, using a single line of wire, a person at one station can send a signal to a distant station.

2. Draw and describe *the commutator* used in a single needle instrument.

3. Explain the use of a *relay*.

4. Describe an electric bell.

MISCELLANEOUS EXAMPLES

EXERCISE XL.

MAGNETISM.

1. You have two rods of steel, one magnetised and the other not. What experiments would you perform to distinguish the one from the other?

2. One pole of a powerful magnet repels the similar pole of a feeble magnet when it is held some distance away. If, however, they are brought close together, attraction takes place. Why?

3. A glass tube full of steel filings can be magnetised; but it loses its magnetism when the filings are shaken up after magnetisation. Explain this.

4. A bar magnet is held vertically, and a piece of soft iron, nearly as heavy as the magnet can support, is placed on the lower end, which is a north-seeking pole. If the south-seeking pole of another magnet be brought under the ball, it holds on more firmly; but if a north-seeking pole be brought below, it falls off. Explain these results.

5. Define accurately the *pole* of a magnet. What is meant by the *strength* of a pole?

6. A bar of very soft iron is set vertically; how will its upper and lower ends respectively affect a compass needle? Would the result be the same at all parts of the world as it is in this country? If not, state generally how it would differ at different places.

7. Two equal and equally magnetised needles are thrust through a wooden match at right angles to it, and with the similar poles pointing in opposite directions. If, when the match is hung by one end from a silk fibre, the needles are not in the same vertical plane, will the combination be astatic? If not, how will it set itself?

8. Give a diagram to illustrate the direction which a magnetic needle, suspended at its centre of gravity, will take under the influence of the earth in our latitude.

9. A magnetic needle, balanced on a pivot so as to move horizontally, makes 11 vibrations in 2 min. 1 sec. at a place A, and 12 vibrations in 2 min. at a place B. Compare the strength of the earth's horizontal magnetic force at A and B.   *Ans.*: 100 : 121.

10. What force does a magnet-pole, the strength of which is 25 units, exert upon a pole the strength of which is 10 units placed 10 centimetres apart?   *Ans.*: 2·5 dynes.

Exercise XLI.

FRICTIONAL ELECTRICITY.

1. A metal ball hung by a silk thread is positively electrified and is then lowered into a hollow insulated conductor so that it is entirely surrounded. How would you prove that the negative charge induced on the interior of the conductor is equal to that on the ball?

2. A Leyden jar is charged and then placed on a sheet of vulcanite. A person touches the knob and outer coating alternately. What happens, and why?

3. An elongated conductor is placed with one end near an insulated and positively charged brass sphere. How would you investigate the electrical condition of the conductor, and what will it be (*a*) when the conductor is insulated, and (*b*) after it is touched with the finger?

4. Give experiments to prove that electricity is discharged from pointed conductors.

5. A large strongly electrified metal ball is brought towards a similar unelectrified ball supported by a dry glass stem as near to it as possible without a spark passing between the balls. The balls remaining at this distance, the unelectrified one is touched with the finger, and immediately there is a spark between it and the other ball. Explain this.

6. Two insulated brass plates, a good way apart, are connected by separate wires with a gold-leaf electroscope, and the electroscope and brass plates are electrified so that there is a small divergence of the gold leaves. How and why is the divergence of the leaves altered when the plates are brought near together and facing each other?

7. An electrified brass plate held over the cap of an electroscope causes the leaves to diverge. On touching the electroscope the leaves fall together. If, after removing the finger, an unelectrified dry glass plate is put between the electrified metal plate and the electroscope, without touching either, the leaves diverge again. Why is this? How must the electrified plate be moved to make the leaves collapse again?

8. Two dry glass bottles stand on the caps of two gold-leaf electroscopes, one upon each electroscope. When mercury is poured into one of the bottles, no divergence of the gold leaves takes place; but the leaves of both electroscopes diverge when the mercury is poured from the first bottle into the second, and the first bottle is then replaced upon its electroscope. Explain this action.

9. A sphere A has a certain quantity of positive electricity, and another sphere B has 10 units of positive electricity; their centres are 2 centimetres apart, and the repulsive force between them is 20 dynes. What is the quantity of electricity on A? *Ans.*: 8 units.

10. Twenty-one units of electricity charge a sphere to a potential 6. What is the radius? *Ans.*: 3·5 cms.

Exercise XLII.

VOLTAIC ELECTRICITY.

1. State exactly what happens when a piece of zinc touches a piece of platinum in dilute sulphuric acid.

2. Give experimental proofs that a voltaic cell and an electrical machine generate electricity of the same nature.

3. Describe the construction of any 'constant cell,' giving the reasons for the employment of the several parts.

4. What do you understand by the strength of a current in a voltaic circuit? Upon what conditions of the battery, and of the external conductor, does the strength of a current depend, and how?

5. State Joule's law respecting the thermal effects of a current of electricity.

6. A metal ring, through which a current of electricity circulates, can move horizontally, its plane remaining always vertical. Describe and explain what happens when one pole or the other of a bar magnet is presented to the ring.

7. What is meant by the 'voltaic arc'? Describe how it is produced.

8. Describe how you would make a right-handed helix. After making it, you place a soft iron rod in it. Explain fully the condition of the rod when a current of electricity from a Grove's battery passes through the helix.

9. A battery has an E.M.F. of 30 volts and its total resistance is 40 ohms. What is the strength of the current?   *Ans.* : ·75 ampère.

10. Explain, with equations, what happens when potassium iodide, copper sulphate, and sodium sulphate are decomposed by the electric current. Tabulate the substances given off at the electrodes.

SOUTH KENSINGTON EXAMINATION PAPER, 1889.

Examiners, Professor A. W. Reinold, F.R.S., and Professor A. W. Rücker, F.R.S.

First Stage or Elementary Examination.

You are not permitted to attempt more than *eight* questions.

You may select only *two* in Magnetism, *three* in Frictional Electricity, and *three* in Voltaic Electricity.

The value attached to each question is the same.

Magnetism.

1. A bar of soft iron, AB, is placed horizontally east and west, the east end, A, being about four inches to the west of the N-seeking pole of a compass-

needle. The end A being fixed, B is raised until the bar is vertical. How is the needle affected by the bar when in its original and final positions ?

2. The beam of a balance is made of soft iron. When it is placed at right angles to the magnetic meridian two equal weights placed in the opposite pans just balance. Will the weights still appear to be equal when the balance is turned so that the beam swings in the magnetic meridian? Give reasons for your answer.

3. An iron ball is held over a pole of a horse-shoe magnet. Will the attraction exerted on the ball be altered if the poles of the magnet are connected by a soft iron keeper, and, if so, in what way, and why?

4. Three precisely similar magnets are placed vertically with their lower ends on a horizontal table. Iron filings are scattered over a plate of glass which rests on their upper ends, two of which are north poles and the third a south pole. Give a diagram showing the forms of the lines of force mapped out by the filings.

Frictional Electricity.

5. An electrified metal ball is introduced into a dry glass tube closed at one end, and then, the tube being held in the hand, is brought near to the cap of an electroscope. What will the effect on the electroscope be if the exterior of the tube (1) is, (2) is not, covered with tinfoil ?

6. The extremity, B, of a wire, AB, is attached to the plate of a gold-leaf electroscope. By means of an insulating handle, the other end, A, is placed in contact first with the blunt and then with the more pointed end of a pear-shaped insulated and electrified conductor.
Describe and explain the movements of the leaves of the electroscope.

7. You charge a Leyden jar by holding its outer coating in the hand and bringing the knob to the prime conductor of an electrical machine. If you wished to charge the jar to as high a potential as possible, would you hold the knob in contact with the prime conductor, or keep them a small distance apart? Give reasons for your answer.

8. A muslin bag containing sulphur and red lead finely powdered is suspended by a silk ribbon so that it hangs within a metal vessel which stands on the cap of an electroscope. When the bag is jerked the powders are shaken out through the muslin into the vessel and become electrified by friction. State and explain what effect, if any, is produced upon the electroscope.

9. Sparks pass between the prime conductor of an electrical machine and a metal knob, connected with the earth, held near to it. Describe the changes, if any, in the phenomena observed, as the knob is gradually moved away from the prime conductor.

Voltaic Electricity.

10. A number of cells formed of plates of zinc and platinum, immersed in dilute sulphuric acid, are to be connected in a circuit, so that the platinum

of each cell is in contact with the zinc of the next. What effect, if any,
would be produced on the current if, by mistake, one cell was made up
with two platinums, instead of with one platinum and one zinc plate?

11. A wire lies east and west (magnetic) immediately over a compass-
needle. How is the direction in which the needle points affected when a
strong current flows through the wire (1) from west to east; (2) from east
to west?

12. Why is an astatic galvanometer better adapted for the measurement
of weak currents than a galvanometer with a single needle?

13. Two Grove's cells, alike in all respects except that in one the plates
are twice as far apart as in the other, are arranged in series, and the poles
of the battery so constituted are united by a copper wire. The liquid in
both cells becomes heated. In which is the rise in temperature the greater,
and why?

14. A guttapercha-covered copper wire is wound round a wooden
cylinder AB from A to B. How would you wind it back from B to A (1)
so as to increase, (2) so as to diminish the magnetic effects which it produces when a current is passed through it? Illustrate your answer by
a diagram drawn on the assumption that you are looking at the end B.

INDEX

———◆———

PRINTED BY
SPOTTISWOODE AND CO., NEW-STREET SQUARE
LONDON

www.ingramcontent.com/pod-product-compliance
Lightning Source LLC
Chambersburg PA
CBHW021523210326

41599CB00012B/1364